Site Assessment and Remediation for Environmental Engineers

Fundamentals of Environmental Engineering
Series Editor
Jeff Kuo

Air Pollution Control Engineering for Environmental Engineers
Jeff Kuo

Site Assessment and Remediation for Environmental Engineers
Cristiane Q. Surbeck and Jeff Kuo

For more information about this series, please visit: https://www.routledge.com/ Fundamentals-of-Environmental-Engineering/book-series/CRCFUNOFENV

Site Assessment and Remediation for Environmental Engineers

Cristiane Q. Surbeck and Jeff Kuo

CRC Press
Taylor & Francis Group
Boca Raton London New York

CRC Press is an imprint of the
Taylor & Francis Group, an **informa** business

First edition published [2021
by CRC Press
6000 Broken Sound Parkway NW, Suite 300, Boca Raton, FL 33487-2742

and by CRC Press
2 Park Square, Milton Park, Abingdon, Oxon, OX14 4RN

© 2021 Taylor & Francis Group, LLC
CRC Press is an imprint of Taylor & Francis Group, LLC

The right of Cristiane Q. Surbeck and Jeff Kuo to be identified as authors of this work has been asserted by them in accordance with sections 77 and 78 of the Copyright, Designs and Patents Act 1988.

Library of Congress Cataloging-in-Publication Data

Names: Surbeck, Cristiane Q., author. | Kuo, Jeff, author.
Title: Site assessment and remediation for environmental engineers /
Cristiane Q. Surbeck and Jeff Kuo.
Description: First edition. | Boca Raton, FL : CRC Press/Taylor & Francis
Group, LLC, 2021. | Series: Fundamentals of environmental engineering |
Includes bibliographical references and index.
Identifiers: LCCN 2020041704 (print) | LCCN 2020041705 (ebook) | ISBN
9781138385450 (hardback) | ISBN 9780429427107 (ebook)
Subjects: LCSH: Hazardous wastes--Risk assessment. | Hazardous waste site
remediation. | Hazardous waste sites--Evaluation.
Classification: LCC TD1050.R57 S93 2021 (print) | LCC TD1050.R57 (ebook)
| DDC 628.4/2--dc23
LC record available at https://lccn.loc.gov/2020041704
LC ebook record available at https://lccn.loc.gov/2020041705

ISBN: 978-1-138-38545-0 (hbk)
ISBN: 978-0-367-70973-0 (pbk)
ISBN: 978-0-429-42710-7 (ebk)

Typeset in Times
by Deanta Global Publishing Services, Chennai, India

Access the Support Material: routledge.com/9781138385450

Dedication

---

*To those affected by contaminated sites and
to those striving to clean them up.*

Contents

Preface

We wrote this book to provide university students, early career professionals, and instructors and professors with a framework to learn and teach about investigating and cleaning up contaminated soil and groundwater. With over 1,000 Superfund sites in the United States and thousands of other contaminated sites with different funding schemes, plus many times these amounts of contaminated sites around the world, skills needed to remediate these sites are sorely necessary.

This book provides the reader with project-oriented knowledge. It focuses on practical calculations and skills needed to investigate a site and design and operate remediation systems. We aim to educate both students and professionals, including consultants, industrial environmental health and safety personnel, and government regulatory employees. The many example problems are followed by brief discussions that review concepts, point out tricky situations, and interpret the results. Using the content of this book, the reader should be able to develop a design and operation and maintenance strategy to clean up a contaminated site.

The first four chapters of the book provide background and needed knowledge on how to investigate a site before proposing how to remediate it. Chapter 1 provides an introduction and context on the importance of studying this subject. Chapter 2 describes the important chemical and soil properties that must be known to assess and remediate a site. Many concepts in later chapters rely on understanding this chapter thoroughly. Chapter 3 describes the primary regulatory framework in the United States for contaminated sites and introduces human health risk assessment. Chapter 4 introduces how to investigate the vadose zone soil and groundwater at a site, in addition to providing basic concepts of groundwater flow.

The next three chapters describe remediation technologies and techniques. Chapter 5 describes vadose zone remediation, and Chapter 6 groundwater remediation. Chapter 7 describes off-gas treatment that is often necessary with some of the technologies employed in Chapters 5 and 6. An instructor may choose to cover parts of Chapter 7 as he/she covers Chapters 5 and 6.

The last two chapters focus on remediation in the long term. Chapter 8 covers performance monitoring and operation and maintenance (O&M), primarily by focusing on the types of data that are acquired in the field and how to analyze them to assess the effectiveness of the remedial actions. Chapter 9 combines the knowledge from the previous chapters to discuss overall remediation strategies and sustainable remediation.

Much of the material can be covered in a one-semester course. Instructors may opt to leave out some material and add supplementary material. For example, instructors may choose to provide more background on the extensive theories of groundwater hydrology, fluid mechanics, chemistry, and microbiology; equation derivations; sustainable remediation; state-of-the-art research; and additional important chemicals of concern. The book will primarily be helpful to engineering (environmental, civil, chemical, geological, and mechanical) and geology students. Other students who can highly benefit from this content come from the fields of environmental science, chemistry, toxicology, and environmental policy.

In writing this book, we made some difficult decisions to leave out certain topics and use a balance of SI and USCS units, often for the sake of reaching a broad audience without overwhelming it. We welcome your feedback and suggestions.

Acknowledgments

This book would not be possible without my coauthor Jeff Kuo, whose reassurance and good humor ("don't worry, be happy") made the writing process fruitful. His earlier work was the invaluable basis for this book.

Several people had a big impact on the outcome of this book. Michael Shiang was an early collaborator and contributed dozens of photos that provide real-life context to the applications in this book. Stefanie Goodwiller's elegant artwork shows up frequently and adds essential visualization to the concepts in the book. Grace Rushing carefully proofread early drafts and asked questions from the point-of-view of a student. She also solved many of the end-of-chapter problems.

My coworkers from my early environmental consulting career in California taught me many of the concepts, applications, and formulas in this book. I credit them with teaching me how to navigate contaminated sites, problem-solving, and professional life.

My present and past students in Mississippi inspired parts of this book. I base some of the figures on their creative artwork and many paragraphs on my answers to the questions they ask.

Many current colleagues and mentors have encouraged me through this process: Richard McCuen, Gregg Davidson, Dwight Waddell, Deb Wenger, the "CE Writing Club," and many others. The Academic Ladder Writing Club was instrumental in teaching me good habits that every writer should have. Working for an academic institution made it possible for me to have part of my schedule dedicated to scholarly work. Thank you to Yacoub Najjar, Dave Puleo, and Alex Cheng at the University of Mississippi for approving my semester sabbatical and course releases.

Mãe and *pai* kept me on my toes, often asking about my progress and making kind suggestions. *Obrigada*! My extended family cheered me from afar.

Greg and Collin were with me every day, patiently experiencing my productive writing days and my messy writing days, plus the seven-day workweeks in the final stretch. They were the best company a wife and mother could have during the uneasy months of the COVID-19 pandemic. Greg brought me tea when I needed energy, and Collin played his piano compositions beautifully while I worked. And they made me laugh every day.

-**Cris Surbeck**

Biographies

Dr. Cristiane (Cris) J.Q. Surbeck is a licensed professional engineer, associate dean, and faculty member of the School of Engineering at the University of Mississippi. She has worked in academia and environmental consulting. In the early part of her career in environmental consulting, her work included investigation, feasibility studies, design, construction, monitoring, and operation and maintenance of cleanup operations of soil and groundwater at contaminated sites; environmental due diligence assessment for manufacturing facilities; and stormwater monitoring programs, with projects throughout the United States, Brazil, and Mexico. At UM, she teaches environmental and water resources engineering courses. Her research topics have ranged from microbial pollutant transport in water bodies, statistical methods for predicting pollutant occurrence, modeling of stormwater quantity and quality using green infrastructure, and infrastructure sustainability. She led the university's chapter of Engineers Without Borders on three trips to work on construction and well drilling projects in a rural village in Togo, West Africa. In 2015, she was elected to the Governing Board of the American Society of Civil Engineers' Environmental and Water Resources Institute (ASCE-EWRI) and served as its president from 2017 to 2018. She received a B.S. degree in civil engineering from the University of Maryland, and an M.S. and Ph.D. in environmental engineering from the University of California, Irvine.

Dr. Jeff (Jih-Fen) Kuo worked in environmental engineering industries for over ten years before joining the Department of Civil and Environmental Engineering at California State University, Fullerton, in 1995. His areas of research in environmental engineering include dechlorination of halogenated aromatics by ultrasound, fines/bacteria migration through porous media, biodegradability of heavy hydrocarbons, surface properties of composite mineral oxides, kinetics of activated carbon adsorption, wastewater filtration, THM formation potential of ion exchange resins, UV disinfection, sequential chlorination, nitrification/denitrification, removal of target compounds using nanoparticles, persulfate oxidation of persistent chemicals, microwave oxidation for wastewater treatment, landfill gas recovery and utilization, greenhouse gases control technologies, fugitive methane emissions from the gas industry, and stormwater runoff treatment. He received a B.S. degree in chemical engineering from National Taiwan University, an M.S. in chemical engineering from the University of Wyoming, an M.S. in petroleum engineering, and an M.S. and a Ph.D. in environmental engineering from the University of Southern California. He is a professional civil, mechanical, and chemical engineer registered in California.

1 Introduction to Assessment and Remediation of Contaminated Sites

In the course of human history, technology has been developed to meet specific purposes. Our lives today are dependent on technologies developed over centuries and especially since the industrial revolution. Electricity, fuels, processed foods, medicines, and dry cleaning of clothes are just a few examples of activities and products of modern life. One thing that all these products and activities have in common is the use of chemicals.

People use chemicals in all aspects of modern life: to fuel our cars and to manufacture clothing, paper products, airplane parts, and home appliances. In the manufacturing process, chemicals are used as part of the product itself, for example, polycarbonate plastic to make the visor for a motorcycle helmet, and as an intermediate process, for example, tetrachloroethylene (PCE) to clean grease off of manufacturing equipment.

While chemicals are useful for all of these activities, when they are misused or improperly disposed of, they become contaminants that affect human health and the environment. For example, chemicals improperly disposed of in a municipal landfill can contaminate groundwater and a city's water well; chemicals leaked in the soil of a neighborhood gas station can volatilize and contaminate homes with noxious fumes.

Contamination cases in the United States in the 1970s and 1980s propelled new laws to deal with these scenarios, and along with them, new applications of science and engineering to assess and remediate the situation. Well-documented cases in the United States include Love Canal in Niagara Falls, NY, Times Beach, MO, and Woburn, MA. Some real cases of soil and groundwater contamination, assessment, and remediation have had detailed accounts described in the page-turner books *A Civil Action* (1995) by Jonathan Harr, winner of the National Book Critics Circle Award for Nonfiction, and *Amity and Prosperity* (2018) by Pulitzer Prize winner Eliza Griswold. Harr's book was popularized by the movie *A Civil Action* (Zaillian 1998), starring John Travolta. Other movies depicting soil and groundwater contamination are *Erin Brokovich* (Soderbergh 2000), starring Julia Roberts, and *Dark Waters* (Haynes 2019), starring Mark Ruffalo.

Woburn, Massachusetts, USA, is the industrial town portrayed in *A Civil Action*. In 1979, after years of residents complaining that their drinking water tasted like chemicals, it was discovered that wells G and H had been contaminated with trichloroethylene (TCE). The wells were shut down, and the city was provided with an alternate source of water. Epidemiological studies determined that Woburn was a cancer cluster, likely due to the residents' ingestion of TCE in their drinking water. Environmental investigations revealed that several industrial facilities were improperly disposing of TCE and other common industrial chemicals. For years, these chemicals leaked into the ground, spreading downward and laterally, contaminating groundwater and the nearby Aberjona River. Site cleanup activities have included excavation, soil vapor extraction, and groundwater pump-and-treat, some of the techniques covered in this textbook. To date, 2,100 cubic yards (1,600 m^3) of contaminated soil have been excavated, 4,700 pounds (2,100 kg) of chemicals have been removed from the soil, and 540 million gallons (2 million m^3) of contaminated groundwater have been pumped (U.S. EPA 2020). Remediation continues at the time of this writing, indicating how severe and widespread contamination problems can be.

The contamination case in the towns of Amity and Prosperity in rural Washington County, Pennsylvania, USA, depicted in the book of the same name (Griswold 2018), are more recent and unresolved. Residents of farms adjacent to a hydraulic fracturing (fracking) operation started noticing foul odors and a bad taste in their water around 2009 and 2010. Farm animals were sick and dying, and children were ill, eventually diagnosed with arsenic poisoning. Air, soil, and groundwater contamination have come from a fracking waste pond uphill of the farms. Fracking is an operation in which deep wells are drilled to extract natural gas for energy. To extract the gas deep in the earth, liquid chemicals are injected into the wells ("hydraulic") to fracture the shale rock ("fracturing") that harbors the gas. What comes out of the wells is a combination of gas, liquids buried deep in the earth, and a portion of the liquid chemicals that were previously injected. When these fluids resurface, they are placed in a constructed waste pond on site. Waste ponds are engineered, designed, and constructed to contain chemicals. However, chemicals leaked through cracks on the bottom of this waste pond and moved through the soil downhill into the neighboring farms. The fracking company was sued, and its environmental managers were implicated. Most residents have since moved out, and environmental investigations are ongoing.

The accounts in these two books focus on the victims of contamination, and rightly so. There is also a heavy focus on the legal actions and the implications for the residents' and attorneys' lives. This textbook focuses on an aspect that happens in the background of these stories: the work by geologists, engineers, and environmental scientists to investigate and clean up contaminated sites.

This book, therefore, addresses assessment and remediation of situations when chemicals leak into the ground, contaminating soil and groundwater. It does so from the perspective of a professional who might be tasked to work on a contamination project from start to finish. The remediation field is interdisciplinary and requires skills from engineering, geology, hydrology, chemistry, microbiology, toxicology, and epidemiology. From an environmental engineering perspective, groundwater and soil remediation has some complications different than conventional drinking water and wastewater treatment. This is because the characteristics of subsurface soil and groundwater, which greatly affect the implementability and effectiveness of a given technology, can only be known by interpolation of a limited number of soil and groundwater samples. Therefore, the selection of a proper remediation strategy is site-specific. One needs to know the applicability and limitations of each technology before an effective decision can be made. In addition to knowing how a remedial technology works, it is essential to know why it may not work for an impacted site.

This book was written for upper-level undergraduate students, graduate students, and early career professionals. It not only conveys the basic principles of site assessment and remediation, but also engages the reader with the practical steps to be taken from first investigating a contaminated site, to selecting appropriate remedial technologies, to operating a remediation system for the long term. This book can serve as a reference book for those employed in industry, consulting companies, law firms, and regulatory agencies in the field of soil and groundwater remediation.

PROBLEMS AND ACTIVITIES

1.1.   Conduct an internet search and write a short essay about the following real-life contamination scenarios, including (a) the chemicals released, (b) what media was contaminated, (c) what health problems surfaced, and (d) the outcome of the case.
   a) Love Canal, New York, USA
   b) Times Beach, Missouri, USA
   c) Woburn, Massachusetts, USA

1.2.   Visit the U.S. Environmental Protection Agency's webpage "Cleanups in My Community" (https://www.epa.gov/cleanups/cleanups-my-community at the time of this writing) and explore your state by clicking on the map and selecting a site. Write a paragraph with information on the site including, for example, the facility name, the chemicals released, the status of human exposure to the chemical, the status of the investigation, and the status of the cleanup.

REFERENCES

Griswold, E. (2018). *Amity and Prosperity: One Family and the Fracturing of America.* Farrar, Straus and Giroux, New York.

Harr, J. (1995). *A Civil Action.* Random House, New York.

Haynes, T. (2019). *Dark Waters.* Focus Features, U.S.A.

Soderbergh, S. (2000). *Erin Brockovich.* Universal Pictures, U.S.A.

U.S. EPA. (2020). *Superfund Site: Wells G & H, Woburn, MA.* U.S. Environmental Protection Agency, https://cumulis.epa.gov/supercpad/SiteProfiles/index.cfm?fuseaction=second. Cleanup&id=0100749#bkground (Aug. 6, 2020).

Zaillian, S. (1998). *A Civil Action.* Buena Vista Pictures, U.S.A.

2 Common Properties of Chemicals of Concern and Soil Matrices

2.1 INTRODUCTION

When a site is suspected to be impacted by chemicals, many technical, regulatory, and legal issues need to be considered to establish what to do next. An environmental practitioner needs to acquire a variety of technical data, for example, historical data on chemical usage and facility operations, identification of source areas, and characteristics of subsurface soil. These data are needed for the development of a comprehensive site assessment plan and a remediation plan if needed. Getting a good handle on the characteristics of chemicals of concern (COCs) and soils is critical for site assessment and subsequent remediation.

Rather than using the term "contaminant," environmental professionals who deal with site investigation and remediation often use the term "chemicals of concern." This is because while chemicals released into the environment are of concern, some are present in low enough concentrations or have low enough toxicity that would not be deemed as a contaminant. The term "COCs" will be used throughout this book and, for variety and ease of reading, may be interchanged with the terms "chemical" or "contaminant."

Understanding the properties of COCs and their behaviors in the subsurface is important. The potential hazards of the COCs determine the urgency of addressing potential contamination. The pathway of a COC from the subsurface to a receptor (such as a human or vegetation) varies in accordance with properties (e.g., solubility and volatility) and concentrations of the COC and the amount of leakage/spill, in addition to the types of media (i.e., water, soil, and air) through which it travels and their characteristics. For example, clayey soil may adsorb more organic chemicals than sandy soil. By having a firm understanding of these COCs and the media through which they travel, one can evaluate, for example, how fast the COC is spreading in the subsurface and the optimal way to remediate the problem.

This chapter will introduce these essential parameters and principles, and they will often be referred to throughout the book for a reader to develop a concise understanding of a site's condition.

2.2 COMMON SOURCES OF POLLUTION

Pollutants in the environment come from many sources. Think of pollutants as chemicals that are toxic to plants, animals, and humans by being present at elevated concentrations where they should not be. Sources can generally be classified into point and nonpoint sources of pollution.

Point sources have well-defined locations. For example, a pipe from an industrial facility discharging wastewater into a stream is a point source to that receiving water body. A stack (chimney) of an industrial facility is a point source with regard to air pollution. A storage tank leaking chemicals into the underlying soil and groundwater is a point source of pollution.

Nonpoint sources of pollution originate from spread-out locations and result in diffuse, or scattered, pollution. For example, stormwater runoff over a developed area carrying various pollutants to a river is a nonpoint source of pollution to that river. Vehicles traveling throughout a city and their tail-pipe emissions constitute a nonpoint source of pollution to the atmosphere. Leachates from pesticides spread over a large crop field constitute a nonpoint source to the underlying soil and ground-water aquifer.

In this section, we will list typical sources of pollution to the subsurface (i.e., soil and groundwater), along with associated COCs. This will help understand how humans are exposed to COCs. To that end, many important questions can be answered after a remediation professional evaluates the source of pollution and the types of chemicals present. How far has the chemical moved in the subsurface? Will the chemical reach a drinking water source? Will the chemical in the subsurface become airborne and rise into an elementary school building through cracks in its foundation? Consequently, will the chemical be a threat to human health and the environment?

Let's first look at typical sources of pollution to the subsurface. The chemical sources, for the purposes of this discussion, are those that potentially occur from leaking chemical storage tanks, leachate from landfills, infiltration of pesticides from agricultural fields, and accidental spills.

2.2.1 LEAKING CHEMICAL STORAGE TANKS

Chemical storage tanks are used in modern human activities. Many of those are installed underground. These underground storage tanks (USTs) are present in fueling stations (e.g., gas stations) and industrial facilities that store chemicals for manufacturing. Historically, USTs for storing hazardous chemicals are often preferable to aboveground storage tanks (ASTs), primarily due to safety concerns. For example, flammable chemicals stored underground pose a smaller fire risk than if stored aboveground. ASTs are also a potential source of contaminants to the subsurface, but their spills are easier to detect and contain, so contamination can be controlled more quickly. However, many ASTs in agricultural settings that store pesticides and fertilizers have been sources of widespread contamination.

Single-walled steel USTs were primarily used until the 1990s, and these tanks were prone to leaks due to cracks and pitting resulting from corrosion. USTs that cracked

FIGURE 2.1   A leaking underground storage tank being pulled from an excavation. (Source: MDEQ 2020.)

and leaked were point sources of pollution that historically were difficult to monitor and detect. They could have leaked for many years before the contamination was discovered. This long-term leakage can heavily contaminate soil and underlying groundwater. Figure 2.1 shows the excavation and removal of a leaking UST, typical of many gas stations where such leaking underground storage tanks (LUSTs) are found.

The U.S. Environmental Protection Agency (U.S. EPA) issued the first regulations on USTs in 1988, with updates in 2015. States can also issue more stringent regulations. For example, in September 2014, the state of California enacted Health and Safety Code Section 25292.05, requiring the permanent closure of all single-walled USTs. The majority of the USTs in the state have been replaced with double-walled, well-coated steel or fiberglass composite tanks, with active vapor and leak detection monitoring systems. Figure 2.2 is a schematic of a modern UST with leakage controls and monitoring systems, such as a line leak detection instrument, a secondary wall to provide secondary containment (a tank outside of the tank, with a monitoring system to detect leaks in the interstitial space), and vapor and groundwater monitoring wells.

2.2.2   LANDFILLS

Another source of chemicals migrating into the subsurface is leachate from landfills. Landfills are engineered earthen structures designed to store solid waste for decades.

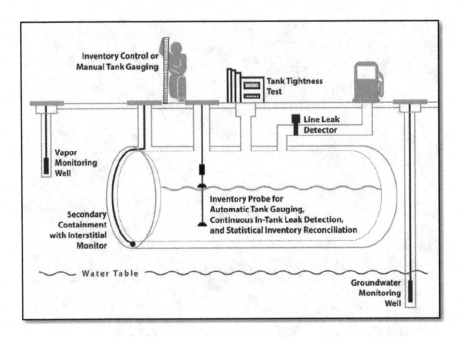

FIGURE 2.2 A modern double-walled UST with leak detection and monitoring systems. (Source: U.S. EPA 2016a.)

A municipal (or sanitary) landfill stores common solid waste from households and businesses. Landfill leachate is caused principally by the inevitable rainfall on landfills that percolates through waste deposited in the landfill. Once in contact with the decomposing solid waste, as well as the liquid it contains, or that is generated from decomposition, the percolating water becomes contaminated. The composition of a landfill leachate depends on the age of the landfill and the waste it contains. With time, leachate may travel downward and reach the leachate collection system, shown in Figure 2.3. The leachate collection system is a characteristic of modern landfills, coupled with an impermeable bottom liner system. However, in landfills that are inadequately designed or constructed, leachate can leak beneath the landfill and contaminate soil and groundwater. A hazardous waste landfill would have more leachate collection and tighter bottom liner systems but is still prone to leaking leachate into soil and groundwater.

2.2.3 Pesticides and Other Agricultural Chemicals

The agricultural industry applies fertilizers, pesticides, and herbicides to crops in fields. Grazing of livestock animals can also be a significant source of contamination due to their fecal matter and waste by-products. These practices have resulted in impacts to the environment, specifically near-surface soil and ultimately surface water and underlying groundwater. Assuming that chemical contaminants eventually make their way into adjacent water bodies via uncontrolled runoff or illicit

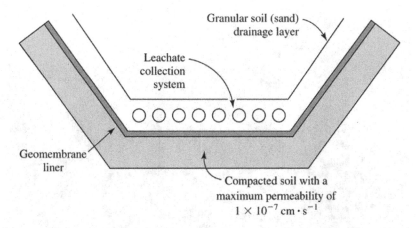

FIGURE 2.3 Cross section of a landfill showing an impermeable geomembrane liner and a leachate collection pipe system. (Source: Davis and Masten 2013, © McGraw-Hill Education.)

discharges from retention basins, an environmental professional could attempt to trace the chemicals back to the origin of contamination, a point source determination. However, if impacts on the environment are ubiquitous and widespread in a groundwater basin, then source identification would be difficult. In such cases, groundwater contaminant plumes from different sources comingle and make it nearly impossible to identify the original chemical source. The contamination then needs to be dealt with regionally instead of locally.

2.2.4 TRANSPORTATION OF CHEMICALS (TANKER TRUCKS AND RAIL CARS)

Another source of contamination is spills from tankers and tank cars. Like ASTs, spills from road and railway accidents are visible and easy to detect. However, they can be very large and difficult to contain and cause catastrophic contamination in a short period, as shown in the tanker spill in Figure 2.4, which spanned several acres. Such events are generally handled with emergency response by hazardous materials (HAZMAT) response teams and require immediate action by local agencies and first responders to assess and control the incident. While immediate abatement can reduce the impact of leaking chemicals to the subsurface, it is the long-term effects of chemicals lost to the ground and seeping into the subsurface that need to be addressed. If a chemical product is lost from a leaking tanker, then further investigation is necessary to evaluate the long-term impacts on soil and underlying groundwater.

2.2.5 ONE-TIME AND CONTINUOUS SOURCES OF POLLUTION

Based on the description of the sources of pollution above, it may be evident that some of them are one-time sources, while some of them are continuous for a certain time. For example, a tanker spill is easily detectable and finite. The spill happens

FIGURE 2.4   A tank car spill in Illinois. (Source: U.S. EPA 2016b.)

over several minutes or hours and ends, at the latest, when the tanker is empty. When a leak or spill has a well-defined beginning and ending time and a relatively short duration, even if it is as long as days or weeks, it can be considered a one-time release. But when a leak from a UST, even if small, is undetected for months or years, and the UST continues to be filled with the chemical while leaking, then that leak is considered a continuous release. An agricultural field where pesticides have been applied and infiltrated into the groundwater for many years is also a continuous source. There is no firm definition for how long a chemical release has to last to be classified as a one-time or continuous source. The classification depends on the situation, but the concept helps to understand how the remediation professional determines whether a source still needs to be eliminated and where to investigate the extent of the contamination.

With this broad view and knowledge of these typical sources of contamination, environmental professionals can turn their attention to what specific groups of chemicals should be analyzed to conduct a thorough site investigation. Keep in mind that the United States has an expansive list of over 85,000 chemicals used in commerce (Erickson 2017). Only a small percentage of these chemicals have toxicological data that explain their health effects on humans and the environment. Reportedly, there are currently over 2,000 new candidate chemicals being considered for introduction into the market (DTSC 2019), mainly from industrial chemical additives, personal care and hygiene products, pharmaceutical wastes, and nanoparticles. With the ability to detect many of these chemicals used commercially, toxicologists have been able to identify a significant number of "emerging" chemicals that were previously unknown or undetected and have evaluated their effects on animals and humans. While there are regulatory limitations to using emerging chemicals, they could appear on the radar because they are found building up in the environment, resulting in adverse effects on public health and the environment.

Understanding how these emerging chemicals build up in the environment and potentially affect human health is critical. This ultimately leads to one's understanding of what the real impacts are in characterizing a site and how potentially to clean up the site. The next section discusses the major COCs.

2.3 COMMON CHEMICALS OF CONCERN

The following sections discuss the types of COCs often present in impacted soil and groundwater, and Table 2.1 lists these. Table 2.1 also shows some important terminologies: volatile organic compounds (VOCs) and persistent organic pollutants (POPs). VOCs are chemicals that are primarily liquids but easily volatilize or evaporate into a gas at ambient temperatures. A VOC's presence in both the liquid and gas phase at the same temperature makes it challenging to investigate and to treat. A POP is a compound that does not easily decay or transfer to a different phase, rendering it difficult to remove from the environment.

2.3.1 FUEL HYDROCARBONS

Fuel hydrocarbons are chemicals that come from fuels such as gasoline, diesel, and jet fuel. Fuels are made up of mixtures of many compounds. As the word "hydrocarbons" implies, the chemical compounds are made of hydrogen and carbon. Because the molecules contain a chain or ring of carbon atoms and come from once-living organisms, they are also classified as organic chemicals. Some common fuel hydrocarbons are benzene (C_6H_6), toluene (C_7H_8), ethylbenzene ($C_6H_5CH_2CH_3$), and xylenes (in the three isometric forms of o-, m-, and p-xylene, $(CH_3)_2C_6H_4$). Note that both ethylbenzene and xylenes have eight carbons and ten hydrogens in their chemical formulas but arranged in a different molecular structure. These four chemicals are collectively known as BTEX (benzene, toluene, ethylbenzene, and xylene).

Crude oil, the origin of refined fuels, is composed of a mixture of hundreds of chemicals. Because fuels originating from crude oil are composed of mixtures of many compounds and their exact composition is proprietary to the chemical companies that manufacture them, the chemical formulas for fuels are not commonly known unless laboratory forensic tests on samples from contaminated areas are used for determining the chemical signatures. Although BTEX are important indicators for the presence of fuel, they do not completely represent fuels. Another term that can be used to represent fuels is "total petroleum hydrocarbons" (TPH), representing a family of compounds composed primarily of hydrogen and carbon.

2.3.2 FUEL OXYGENATES

Fuel oxygenates are chemicals that are added to gasoline to promote cleaner emissions from vehicles. The fuel oxygenates usually contain alcohols and ethers. Methyl tertiary butyl ether (MTBE, $(CH_3)_3COCH_3$), tertiary butyl alcohol (TBA, $(CH_3)_3COH$), ethyl tertiary butyl ether (ETBE, $C_2H_5OC(CH_3)_3$), tertiary amyl methyl ether (TAME, $C_6H_{14}O$), and ethanol (ethyl alcohol, EtOH, CH_3CH_2OH) are common

TABLE 2.1

A Listing of Common Chemical of Concern (COC) Classifications and Examples

Classification	Common examples
Fuel hydrocarbons	Benzene, toluene, ethylbenzene, xylenes (BTEX)
	Total petroleum hydrocarbons (TPH)
	•Diesel
	Many fuel hydrocarbons are volatile organic compounds (VOCs)
Fuel oxygenates	Methyl tertiary butyl ether (MTBE)
	Ethanol
Chlorinated solvents	Tetrachloroethylene (PCE)
	Trichloroethylene (TCE)
	Many chlorinated solvents are VOCs
Pesticides	Alachlor
	Atrazine
	Dichloro-diphenyl-trichloroethane (DDT)
	Many pesticides are chlorinated chemicals (but not necessarily solvents) *and are considered to be persistent organic pollutants (POPs)*
Polychlorinated biphenyls (PCBs)	Aroclor
	Askarel
	Also considered to be POPs
Per- and polyfluoroalkyl substances (PFAS)	Perfluorooctanoic acid (PFOA)
	Perfluorooctane sulfonic acid (PFOS)
	Also considered to be POPs
Heavy metals	Cadmium
	Chromium
	Lead
	Mercury
Dioxins and furans	Polychlorinated dibenzodioxins
	Polychlorinated dibenzofurans
	Also considered to be POPs
Polycyclic aromatic Hydrocarbons (PAHs)	Anthracene
	Naphthalene
	Pyrene

fuel oxygenates found in impacted soil and groundwater because of leaking fuel USTs. In the United States, the reformulation of oxygenated gasoline primarily uses ethanol.

MTBE is an example of an emerging chemical that was used widely without proper knowledge of its effects on the environment. It was introduced in the 1980s but was eventually banned. While MTBE contributed to cleaner air, it became a common groundwater contaminant. It is highly soluble in water, does not biodegrade easily, and is difficult to remove via groundwater treatment. MTBE threatened many drinking water supplies throughout the groundwater basins in regions where it was widely used. Even today after it has been banned for many years, MTBE is still a contaminant present in groundwater throughout the United States.

2.3.3 CHLORINATED SOLVENTS

Chlorinated solvents are synthetic organic chemicals used in industrial processes, either in the manufacturing of goods or for cleaning and removing grease (dissolving grease, hence working as a "solvent") from industrial equipment. They are hydrocarbon molecules that include chlorine atoms. Common chlorinated solvents are tetrachloroethylene (or perchloroethylene, PCE, $Cl_2C=CCl_2$) and trichloroethylene (TCE, $ClCH=CCl_2$). PCE is a common chemical for dry cleaning and is often found in the soil and groundwater beneath commercial establishments that have dry cleaning businesses where spills and leaks have occurred over many years. Because PCE contamination became an extensive problem for dry cleaning facilities, modern facilities have established new and more efficient procedures for handling the chemical. Other common chlorinated solvents of concern include 1,1,1- or 1,1,2-trichloroethane (TCA, $C_2H_3Cl_3$), 1,1- or 1,2-dichloroethane (DCA, $C_2H_4Cl_2$), cis- or trans-1,2-dichloroethylene (DCE, $C_2H_2Cl_2$), and methylene chloride (dichloromethane, CH_2Cl_2).

2.3.4 PESTICIDES

Pesticides are substances that are applied to kill off unwanted pests, such as insects and weeds. Two large categories of pesticides are insecticides (to kill insects) and herbicides (to kill all vegetation or specific weeds and brush). These types of chemicals can be applied to a row crop, a grove of fruit trees, a grassy area along a highway, or a house infested with termites. The application of pesticides is typically a nonpoint source of soil and groundwater pollution. If excessively applied, pesticides can slowly percolate down into the soil and groundwater instead of being uptaken by the insects or plants they were intended to kill. Many COCs in this category are chlorinated organic chemicals. Examples of these COCs are alachlor ($C_{14}H_{20}ClNO_2$), atrazine ($C_8H_{14}ClN_5$), chlordane ($C_{10}H_6Cl_8$), dichloro-diphenyl-trichloroethane (DDT, $C_{14}H_9Cl_5$), 2,4-dichlorophenoxyacetic acid (2,4-D, $C_8H_6Cl_2O_3$), 6-sec-butyl-2,4-dinitrophenol (dinoseb, $C_{10}H_{12}N_2O_5$), endrin ($C_{12}H_8Cl_6O$), heptachlor ($C_{10}H_5Cl_7$), and lindane ($C_6H_6Cl_6$). DDT, chlordane, dinoseb, endrin, and heptachlor have been banned in the United States due to their excessive toxicity. But they still can be found in the subsurface because of decades of use in the past.

2.3.5 POLYCHLORINATED BIPHENYLS

Polychlorinated biphenyls (PCBs) are a class of chemicals composed of chlorine, hydrogen, and carbon atoms. A PCB has two benzene rings in its structure, with a chemical formula of $C_{12}H_{10-x}Cl_x$. PCBs were manufactured mainly as oily liquids to be used as lubricants and coolers in transformers and other electrical equipment. Although the manufacturing of PCBs in the United States ceased in 1977, PCBs have remained in old equipment and in the environment. PCBs are suspected human carcinogens and cause cancer in animals.

2.3.6 Per- and Polyfluoroalkyl Substances

Per- and polyfluoroalkyl substances (PFAS) are synthetic chemicals that repel water or oil. They are often used as a coating in packaging materials that need to stay dry, on materials that can easily get stained, and in nonstick pots and pans. They enter the environment through the manufacturing process and the wear and tear of products that contain them. These chemicals are organofluorine compounds with fluorine atoms attached to an alkyl chain. The most common ones include perfluorooctanoic acid (PFOA) and perfluorooctane sulfonic acid (PFOS).

2.3.7 Heavy Metals

Metals occur naturally in the environment. Those that are toxic, whether at high or low concentrations, are commonly referred to as "heavy" metals in the environmental and toxicology fields, though there is no exact definition for a heavy metal. Metals are often used in industrial processes and end up in soil and groundwater primarily when dissolved in another liquid, such as water or a solvent. Typical heavy metals of concern in the environment are arsenic (technically an element with characteristics of metals and nonmetals), cadmium, chromium, lead, and mercury.

2.3.8 Other Chemicals of Concern

Other COCs are dioxins, furans, and polycyclic aromatic hydrocarbons (PAHs). Although it is impossible to cover all COCs in a single textbook, their characteristics can be found in the literature and used for conducting the same calculations shown in this chapter and others in this textbook. Common literature for information on COCs can be found through the U.S. EPA, Centers for Disease Control and Prevention (CDC) and its Agency for Toxic Substances and Disease Registry (ATSDR), and *Yaws' Handbook of Properties for Aqueous Systems* (Yaws 2012).

2.4 UNITS OF CONCENTRATION AND MASS

Site contamination involves COCs present in water, soil, and air. We will go over units of COC concentration in these media. Think of concentration essentially as a ratio of the amount of COCs per unit mass or unit volume of the media. Environmental professionals need to understand the relationship among mass, volume, and concentration. These dimensions are essential for applying formulas used for site investigation and remediation.

A leaking UST is illustrated in Figure 2.5. The vadose zone (also called the unsaturated zone, or zone of aeration) is the subsurface between the ground surface level (gsl) and the water table. The soil in the vadose zone is not fully saturated with water; in other words, the pores contain both air and water. Below the water table is the phreatic zone, in which the soil is saturated with water (i.e., the aquifer). The water table is the interface between the vadose zone and the uppermost aquifer.

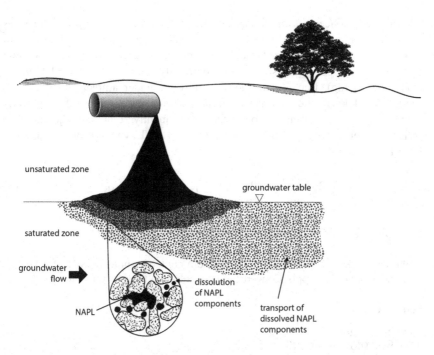

FIGURE 2.5 Partitioning of chemicals leaking from an underground storage tank (UST), moving from the unsaturated (vadose) zone to the groundwater aquifer. NAPL = nonaqueous phase liquid.

As the chemical leaks out of a leaking UST, the liquid will move down and slightly outward. If it is a VOC, such as ethanol or benzene, some of it will volatilize into the air void. Because the soil is usually not completely dry, there is a moisture layer on top of the soil grains. Through contact with the moisture layer, some of the leaked liquid will get dissolved into the soil moisture. Some of the dissolved COCs will then get adsorbed onto the surface of the soil grains. As the leaked liquid continues to infiltrate down, it reaches the aquifer. Some of it will get dissolved into the groundwater, while the remaining liquid will sit on top of the water table. Figure 2.5 also shows that a single chemical can partition (that is, distribute into parts) into and be present in different phases, each with its own concentration. Additionally, the chemical can be present in its undiluted liquid form called the nonaqueous phase liquid (NAPL).

2.4.1 Concentration and Mass of COCs in Liquid

The concentration of a chemical in a liquid is typically expressed in terms of its mass per unit volume of the liquid. For example, 1 mg/L means one milligram (mg) of a chemical in one liter of liquid, while 1 µg/L is one microgram per liter, which is one-thousandth of 1 mg/L.

Alternatively, concentrations can be expressed as the mass of chemical per mass of the liquid or solution, for example, mg/mg. Since concentrations of chemicals in the water in the environment are typically low, one part per million (ppm), meaning 1 mg (of chemical) per kg (of solution) is commonly used (i.e., 1 ppm=1 mg/kg= 1 mg/1,000,000 mg). Because the density of water under ambient conditions is approximately 1 kg/L, the volume of a dilute solution with a mass of 1 kg under ambient conditions is essentially one liter. Consequently, 1 ppm is equivalent to 1 mg/L for water. Similarly, 1 ppb is equivalent to 1 µg/L, which is one-thousandth of 1 mg/L (or 1 ppm).

$$1\frac{\text{mg COC}}{\text{L water}} = 1\frac{\text{mg COC}}{\text{kg water}} = 1\text{ ppm} = 1,000\text{ ppb} \qquad (2.1)$$

Example 2.1: COC Concentrations in Liquid

In the United States, the established maximum contaminant level (MCL) for lead in drinking water is 15 ppb. Convert this concentration to µg/L, ppm, and mg/L.

SOLUTION:

15 ppb=15 µg/L
15 ppb=0.015 ppm=0.015 mg/L

In site assessment and/or remediation design, it is often necessary to determine the mass of a COC present in a medium. That can be found from the COC concentration and the amount of the medium containing the COC. The procedure for such calculations is different for the liquid, solid (soil), and air phases. The differences mainly come from the differences in units of concentration for these three media.

Let us start with the simplest case in which a liquid is impacted with a dissolved COC. As mentioned, the dissolved COC concentration in the liquid (C) is typically expressed in the mass of COC/volume of liquid, such as mg/L; therefore, the mass of the COC dissolved in a liquid can be obtained by multiplying the concentration with the volume of liquid (V_l):

$$\text{Mass of COC in liquid } = (\text{liquid volume})\times(\text{liquid concentration}) = (V_l)(C) \qquad (2.2)$$

2.4.2 Concentration and Mass of COCs in Soil

The concentration of a chemical in the soil is typically expressed in terms of its mass per unit mass of the soil. For example, 1 mg/kg means one milligram (mg) of a chemical in one kg of soil, while 1 µg/kg is one microgram per kg, which is one thousandth of 1 mg/kg. In practical applications, part per million (ppm) is also commonly used; 1 ppm=1 mg/kg=1,000 µg/kg=1,000 ppb.

In environmental investigation and remediation activities, it is typical for workers to collect soil samples in glass jars, as illustrated in Figure 2.6, and send them to a laboratory for chemical analysis. Because soil samples usually contain some

FIGURE 2.6 Collecting soil samples in glass jars. (Source: U.S. EPA 2018.)

moisture in addition to the COC, the water moisture needs to be eliminated before chemical analysis. In the laboratory, the sample would be weighed first. The measured value includes the weight of the dry soil and that of the associated moisture. If the soil is impacted, the COCs should be present on the surface of the soil grains (as adsorbed) as well as in the soil moisture (as dissolved). Both the adsorbed and dissolved COCs in this soil sample would then be extracted and quantified as a whole. The COC concentration in soil (X) would be reported in the unit of "mass of COC/mass of soil," such as mg/kg. The mass of a COC in the soil can be obtained by multiplying its concentration in soil with the mass of soil (M_s):

$$\text{Mass of COC in soil} = (\text{mass of soil}) \times (\text{COC concentration in soil}) = (M_s)(X) \quad (2.3)$$

The mass of the soil can be estimated as the product of the volume and bulk density of the soil. Bulk density is the mass of a material divided by the total volume it occupies. For civil engineering practices, the reported values of the bulk density are often on a dry soil basis, that is, "dry" bulk density (= mass of dry soil ÷ volume as a whole). However, the COC concentration in soil is usually based on the mass of wet soil (soil + moisture). Consequently, the bulk density used to calculate the soil mass for subsequent estimation of COC mass should be the "wet" bulk density (= mass of dry soil plus moisture ÷ volume as a whole). The "wet" bulk density is also referred to as the "total" bulk density. In this book, ρ_t is the symbol for total bulk density and ρ_b is the symbol for (dry) bulk density. When using soil concentrations in calculations, we assume that the chemical is evenly distributed in the soil sample. The mass of COC in the soil can also be found as:

$$\text{Mass of COC in soil} = \left[(\text{soil volume})(\text{total bulk density})\right] \times (\text{COC concentration in soil})$$

$$= \left[(V_s)(\rho_t)\right](X) = (M_s)(X)$$

$$(2.4)$$

2.4.3 CONCENTRATION AND MASS OF COCS IN AIR

Concentrations of chemicals in the air are usually expressed in volume concentration (e.g., ppm), mass concentration (e.g., mg/m³), or partial pressures. The meaning of concentration in ppm in the air is completely different from that in liquid or in soil. One ppm (or ppm by volume, ppmV) of benzene in the air means one part volume of benzene in one million parts volume of air space. Conversions between the mass and volume concentrations are often needed. It can be done by using one of the two equations given below. G is the concentration of the gaseous COC in the air, MW is the molecular weight of the compound, and MV is the molar volume of the air at that temperature and pressure:

$$G\left(\text{in} \frac{\text{mg}}{\text{m}^3}\right) = G(\text{in ppmV})\left(\frac{\text{MW}}{\text{MV(in L)}}\right) \tag{2.5}$$

$$G\left(\text{in} \frac{\text{lb}}{\text{ft}^3}\right) = G(\text{in ppmV})\left(\frac{\text{MW}}{\text{MV(in ft}^3)}\right) \times 10^{-6} \tag{2.6}$$

For typical environmental applications, MV of air can be estimated using the Ideal Gas Law:

$$PV = \left(\frac{m}{\text{MW}}\right)RT = nRT \tag{2.7a}$$

So the molar volume (MV) of air is Eq. 2.7a rearranged with $n = 1$:

$$MV = V = \frac{nRT}{P} = \frac{RT}{P} \tag{2.7b}$$

where P = absolute pressure (atm, kPa, Pa, or psi); V = volume of gas (L, m³, or ft³); m = mass (g, kg, or lbm); n = number of moles of gas; T = absolute temperature (degrees Kelvin or Rankine); and R = ideal gas constant. Below are the values of R in some commonly used units:

$$R = 0.08206 \frac{\text{L} \cdot \text{atm}}{\text{g-mole} \cdot \text{K}} = 8.314 \frac{\text{m}^3 \cdot \text{P}_a}{\text{g-mole} \cdot \text{K}} = 10.731 \frac{\text{ft}^3 \cdot \text{psi}}{\text{lb-mole} \cdot \text{R}}$$

In chemistry, mole refers to g-mole. For example, one mole of water (H_2O) has a molar mass of 18 grams, or 18 g/mol. However, in engineering practice, kg-mole and lb-mole are also commonly used. One kg-mole of water has a mass of 18 kilograms, or 18 kg/kmol, while one lb-mole of water has a mass of 18 pounds.

To understand the units behind Eq. 2.5, the full equation is:

$$G\left(\text{in} \frac{\text{mg}}{\text{m}^3}\right) = G(\text{in ppmV})\left(\frac{\frac{1\,\text{m}^3\,\text{of COC}}{10^6\,\text{m}^3\,\text{of air}}}{\text{ppmV}}\right)\left(\frac{\text{MW of COC}\left(\frac{\text{g}}{\text{mol}}\right)}{\text{V of 1 mole of COC}\left(\frac{\text{m}^3}{\text{mol}}\right)}\right)\left(10^3\frac{\text{mg}}{\text{g}}\right) \tag{2.8}$$

where

G (in mg/m³) = concentration of gaseous COC in the desired units of mg/m³

G (in ppmV) = concentration of gaseous COC in the given unit of ppmV

MW = molecular weight of the gaseous COC, in units of g/mol

V = volume occupied by 1 mole of the gaseous COC, at a given temperature and pressure in units of m³/mol. This is the volume found using the Ideal Gas Law.

The term $\left(\dfrac{\dfrac{1\,m^3\,of\,COC}{10^6\,m^3\,of\,air}}{ppmV} \right)$ is the definition of ppmV, which is used as a conver-

sion factor. The term $\left(10^3\,\dfrac{mg}{g} \right)$ is also a conversion factor.

Example 2.2: COC Concentrations in Air

Find the volume of (a) 1 g-mole and (b) 1 lb-mole of an ideal gas at T = 0°C and P = 1 atm.

SOLUTION:

(a) For 1 g-mole of an ideal gas,

$$V = \frac{nRT}{P} = \left[(1g\text{-mole}) \left(0.08206 \frac{L \cdot atm}{g\text{-mole} \cdot K} \right) (273K) \right] \div 1\,atm = 22.4L$$

(b) For 1 lb-mole of an ideal gas,

$$V = \left[(1\,lb\text{-mole}) \left(10.731 \frac{ft^3 \cdot psi}{lb\text{-mole} \cdot R} \right) (492R) \right] \div 14.7\,psi = 359\,ft^3$$

DISCUSSION:

1. Using similar approaches, one can readily find that volumes of 1 g-mole ideal gas @T = 20 and 25°C (P = 1 atm) are 24.05 and 24.46 L, respectively, while those of 1 lb-mole ideal gas would be 385 and 392 ft³ @T = 68 and 77°F, respectively.
2. It may not be a bad idea to have these values readily available or memorized.

Example 2.3: COC Concentrations in Air

At P = 1 atm, determine the conversion factors between ppmV and mg/m³ or lb/ft³ for (a) benzene (C_6H_6, MW = 78 g/mol) at T = 20°C and 25°C and (b) tetrachloroethylene (PCE, C_2Cl_4, MW = 166 g/mol) at T = 20°C.

SOLUTION:

(a) $1\,\text{ppmV benzene} = \left(\dfrac{78}{24.05}\right)\dfrac{\text{mg}}{\text{m}^3} = 3.24\dfrac{\text{mg}}{\text{m}^3}\,@\,20°C$

$= \left(\dfrac{78}{24.46}\right)\dfrac{\text{mg}}{\text{m}^3} = 3.19\dfrac{\text{mg}}{\text{m}^3}\,@\,25°C$

$1\,\text{ppmV benzene} = \left(\dfrac{78}{385}\right)\times10^{-6}\dfrac{\text{lb}}{\text{ft}^3} = 2.03\times10^{-7}\dfrac{\text{lb}}{\text{ft}^3}\,@\,20°C\,(68°F)$

$= \left(\dfrac{78}{392}\right)\times10^{-6}\dfrac{\text{lb}}{\text{ft}^3} = 1.99\times10^{-7}\dfrac{\text{lb}}{\text{ft}^3}\,@\,25°C\,(77°F)$

(b) $1\,\text{ppmV PCE} = \left(\dfrac{166}{24.05}\right)\dfrac{\text{mg}}{\text{m}^3} = 6.90\dfrac{\text{mg}}{\text{m}^3}\,@\,20°C$

$= \left(\dfrac{166}{385}\right)\times10^{-6}\dfrac{\text{lb}}{\text{ft}^3} = 4.31\times10^{-7}\dfrac{\text{lb}}{\text{ft}^3}\,@\,20°C\,(68°F)$

DISCUSSION:

1. From part (a), we can see that the conversion factor for a compound is temperature-dependent because its molar volume varies with temperature. The higher the temperature is, the smaller the mass concentration would be for the same ppmV value. Please note that the molar volume is also pressure-dependent.
2. From part (b), we can see that the conversion factors are different among compounds because of the differences in molecular weight. The mass concentration of 1 ppmV PCE is twice that of 1 ppmV benzene (6.90 vs. 3.24 mg/m³).
3. Note: 1 lb/ft³ $= 1.603\times10^7$ mg/m³

The mass of a COC in the air can then be obtained by multiplying its mass concentration (G) with the volume of the air (V_a):

$$\text{Mass of COC in air} = (\text{air volume})\times(\text{COC concentration in mass/volume}) = (V_a)(G) \quad (2.9)$$

Example 2.4: Mass and Concentration Relationship (in U.S. customary system (USCS) of units)

Which of the following media contains the largest mass of xylenes $[C_6H_4(CH_3)_2]$?

(a) 1 million gallons of water containing 10 ppm of xylene
(b) 100 cubic yards of soil (total bulk density $= 1.8$ g/cm³) having 10 ppm of xylenes
(c) An empty warehouse (200′ × 50′×20′) containing 10 ppmV xylenes in air (T$=20°C$)

SOLUTION:

(a) Mass of xylenes in liquid = (liquid volume)(liquid concentration)
$$= [(1,000,000 \text{ gallon})(3.785 \text{ L/gallon})](10 \text{ mg/L})$$
$$= (3.785 \times 10^6)(10) = 3.79 \times 10^7 \text{ mg}$$

(b) Mass of xylenes in soil
$$= [(\text{soil volume})(\text{total bulk density})](\text{COC concentration in soil})$$
$$= \{[(100 \text{ yd}^3)(27 \text{ ft}^3/\text{yd}^3)(30.48 \text{ cm/ft})^3] \times [(1.8 \text{ g/cm}^3)(\text{kg/1,000 g})]\}$$
$$(10 \text{ mg/kg})$$
$$= (1.37 \times 10^5)(10) = 1.37 \times 10^6 \text{ mg}$$

(c) MW of xylenes $[C_6H_4(CH_3)_2] = (12)(6) + (1)(4) + [12 + (1)(3)](2) = 106$ g/mole
At $T = 20°C$ and $P = 1$ atm,
   G of xylenes = (10 ppmV)(MW of xylene/24.05) = (10)(106/24.05) = 44.07 mg/m^3
Mass of xylenes in air = (air volume)(vapor concentration)
$$= [(200 \times 50 \times 20 \text{ ft}^3)(0.3048 \text{ m/ft})^3](44.07 \text{ mg/m}^3)$$
$$= (5.66 \times 10^3)(44.07) = 2.5 \times 10^5 \text{ mg}$$
The *water* contains the largest amount of xylene.

DISCUSSION:

1. The concentrations of xylenes are the same in all three media (10 ppm), but the masses are different.
2. To convert a ppmV value in air to a mass concentration basis, the corresponding temperature and pressure need to be specified (P here is assumed to be 1 atm).

Example 2.5: Mass and Concentration Relationship (in SI units)

Which of the following media contains the largest amount of toluene $[C_6H_5(CH_3)]$?

(a) 5,000 m^3 of water containing 5 ppm of toluene
(b) 5,000 m^3 of soil (total bulk density = 1,800 kg/m^3) having 5 ppm of toluene
(c) An empty warehouse (indoor space = 5,000 m^3) with 5 ppmV toluene in air ($T = 25°C$).

SOLUTION:

(a) Mass of COC in liquid = (liquid volume)(dissolved concentration)
$$= [(5,000 \text{ m}^3)(1,000 \text{ L/m}^3)](5 \text{ mg/L}) = (5 \times 10^6)(5) = 2.5 \times 10^7 \text{ mg}$$

(b) Mass of COC in soil = [(soil volume)(total bulk density)](COC concentration in soil)
$$= [(5,000 \text{ m}^3)(1,800 \text{ kg/m}^3)](5 \text{ mg/kg}) = (9.0 \times 10^6)(5) = 4.5 \times 10^7 \text{ mg}$$

(c) MW of toluene $[C_6H_5(CH_3)] = (12)(7) + (1)(8) = 92$ g/mole
At $T = 25°C$ and $P = 1$ atm,
   G of toluene = (5 ppmV)(MW of toluene/24.5)
$$= (5)(92/24.5) = 18.76 \text{ mg/m}^3$$
Mass of COC in air = (air volume)(vapor concentration)
$$= [5,000 \text{ m}^3](18.76 \text{ mg/m}^3) = 9.38 \times 10^4 \text{ mg}$$
The soil contains the largest amount of toluene.

DISCUSSION:

1. Using SI units appears to be easier in this type of calculation. However, engineers, at least in the United States, need to master unit conversions in their job assignments because the USCS of units is still commonly used in the workplace.
2. With the same volume of 5,000 m³ and the same concentration of 5 ppm, the amounts in these three media are quite different.
3. Be aware that the equations for ppmV to mass concentration conversion are different between this example and Example 2.4 because the temperatures are different (25 vs. 20°C).

2.5 PHYSICAL AND CHEMICAL PROPERTIES OF COCS

2.5.1 DENSITY AND SPECIFIC GRAVITY

The density of a fluid is its mass divided by its volume, or mass per unit volume. For fluids, density varies with temperature and pressure. For example, the density of water at 25°C and 1 atm is 0.997 g/cm³ and at 3.98°C is 1.00 g/cm³ (or 1.00 g/mL). We generally approximate the density of water under ambient conditions to 1 g/cm³, 1,000 kg/m³, or 1 kg/L. But some situations require the use of the exact density associated with a temperature and a pressure.

For site assessment/remediation, NAPLs are often classified into two classes: those that are lighter than water, with densities less than 1 kg/L, are called light NAPLs (LNAPLs), and those with a density greater than water are called dense NAPLs (DNAPLs) (Figure 2.7). Most commonly encountered LNAPLs in site remediation are hydrocarbon fuels such as gasoline, diesel, and jet fuel. Gasoline is a

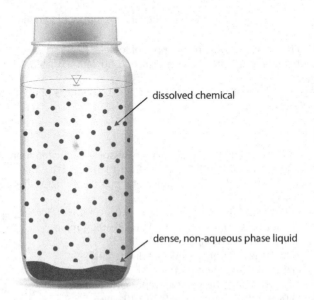

FIGURE 2.7 Illustration of a chemical dissolved in water (circles are homogeneously distributed in the bottle) and as a dense nonaqueous phase liquid (DNAPL) on the bottom.

mixture of many hydrocarbons, and its density is around 0.75 kg/L under ambient conditions. Most DNAPLs encountered are chlorinated hydrocarbons such as dichloroethane (DCA), dichloroethylene (DCE), trichloroethane (TCA), trichloroethylene (TCE), and tetrachloroethylene (PCE). This basic chemical characteristic of density is important to environmental professionals in assessing the chemical's potential movement in the subsurface. LNAPLs will float on the surface of the water table, as also seen previously in Figure 2.5, while DNAPLs will sink to the bottom of the aquifer and sit on the top of the aquitard.

For gases, density varies according to temperature and pressure, as Eq. 2.6 indicates when you consider MW/MV to be equivalent to density (because n can be converted to mass using the molecular weight). Assuming the molecular weight of air is equal to 29, the density of air at 1 atm is 1.18 kg/m^3 at 25°C (MV = 24.5 L) and 1.29 kg/m^3 at 0°C (MV = 22.4 L).

Although the dimensions for density are the same as those for concentration (mass/volume), density is totally different from concentration. Density is mass/volume of the same substance, and concentration is mass/volume of a substance in another substance.

Specific gravity (SG) is the ratio of the density of one substance to the density of a reference substance, usually water. Taking 1.1-dichloroethane (1,1-DCA), a DNAPL, as an example, its density is 1.2 g/mL, and its specific gravity is 1.2 (i.e., 1.2 g/mL ÷ 1.0 g/mL = 1.2).

Example 2.6: Density, Mass, and Concentration

10,000 gallons of water contaminated with trichloroethylene (TCE, density 1.46 kg/L) are stored in an air-tight tank. Several samples of water were collected, and the average concentration of TCE is 200 µg/L (ppb).

(a) What is the total mass of TCE in this tank?
(b) What would be the volume of TCE in its pure liquid form?

SOLUTION:

(a) Mass of TCE in liquid = (liquid volume)(dissolved concentration)
  = [(10,000 gal)(3.785 L/gal)](200 µg/L) = 7.57×10^3 mg = 7.57 g = 7.57×10^{-3} kg
(b) Volume of TCE = Mass ÷ Density
  = 7.57×10^{-3} kg ÷ 1.46 kg/L = 5.18×10^{-3} L = 5.18 mL

Discussion:

The maximum contaminant level, or drinking water standard, for TCE is 5 ppb. Therefore, 200 ppb is a very high concentration. Note that only a small amount (7.57 g; 5.18 mL) of TCE makes 10,000 gallons of water very unsafe to drink.

2.5.2 Solubility

Solubility is the degree that a substance (the solute) will dissolve into another substance (the solvent). Those who like a lot of sugar in their coffee will relate to this.

The first spoonful of sugar will easily dissolve into a hot cup of coffee. The more spoonfuls of sugar are poured into the cup; the more vigorous mixing is necessary for the sugar to dissolve. At some point, more added sugar will not dissolve, and the grains will settle to the bottom of the cup. The concentration of sugar that remains dissolved in the cup is the solubility of sugar in coffee at the given temperature. The degree of solubility of any solute in water is expressed in units of concentration or percentage and depends on the temperature. Although a common saying is that "oil and water don't mix," oil actually does dissolve in water to some extent. If one were to mix oil and water in a cup, the oil will be dissolved in water at a concentration equal to its solubility. The remainder of the oil will float above the water in its original state. Oil will float on water because it is less dense than water, an LNAPL. As a reminder, a substance denser than water will settle to the bottom, as seen in Figure 2.7, even while some of it is dissolved in the water.

Solubility is important for understanding the fate and transport of COCs in the environment. Solubility varies significantly among chemical compounds. For example, alcohols are very soluble in water, and chlorinated compounds are less soluble. If a compound is more soluble, it will be dissolved more readily into soil moisture. In addition, once a very soluble chemical enters the aquifer, it will easily dissolve into the groundwater and travel with the movement of groundwater to form a dissolved plume. However, it should be noted that the solubility value of a compound reported in the literature is the maximum dissolved quantity of that compound in water (at a specified temperature) under laboratory conditions. The actual dissolved concentrations in impacted aquifers are typically much smaller than their corresponding solubility values. For example, the solubility of benzene in water at 25°C is 1.78×10^3 mg/L (Table 2.2). However, the benzene concentrations in groundwater samples collected from aquifers are typically much smaller than this value.

2.5.3 VAPOR PRESSURE AND VOLATILITY

Volatile compounds are those with a high vapor pressure, which also means having a low boiling point. They can be thought of as those liquids that easily evaporate. Water is volatile to some extent, but many chemicals are more volatile than water, such as benzene and gasoline (a mixture of many chemicals), alcohol, and TCE. It is important to know the volatility of chemicals because, though at ambient temperatures they are liquids, their vapors easily volatilize and can be harmful to humans via inhalation. Furthermore, when those chemicals migrate into the subsurface, their vapors will be present in the vadose zone.

Let's visualize a scenario with a common VOC, alcohol. Since alcohol is volatile, the air space above the alcohol inside a container, known as "headspace," will contain alcohol in the vapor phase. That vapor will exert pressure on the surfaces of the headspace (for example, the container and lid), including the air-liquid interface. We can keep this visualization in mind when defining vapor pressure. Vapor pressure is the pressure exerted by a vapor on a liquid at equilibrium in a closed environment. Vapor pressure is highly temperature-dependent. Its units are units of pressure, typically mm of mercury (mm-Hg), atmosphere (atm), kPa, and psi. The vapor pressures

TABLE 2.2
Physicochemical Properties of Common COCs

Compound	MW	H	Pvap	D	Log K$_{ow}$	Solubility	T
	(g/mole)	(atm/M)	(mm-Hg)	(cm²/s)		(mg/L)	(°C)
Benzene	78.1	5.55	95.2	0.092	2.13	1,780	25
Bromomethane	94.9	106	–	0.108	1.10	900	20
2-butanone	72	0.0274	–	–	0.26	268,000	–
Carbon disulfide	76.1	12	260	–	2	2,940	20
Chlorobenzene	112.6	3.72	11.7	0.076	2.84	488	25
Chloroethane	64.5	14.8	–	–	1.54	5,740	25
Chloroform	119.4	3.39	160	0.094	1.97	8,000	20
Chloromethane	50.5	44	349	–	0.95	6,450	20
Dibromochloromethane	208.3	2.08	–	–	2.09	0.2	–
Dibromomethane	173.8	0.998	–	–	–	11,000	–
1,1-Dichloroethane	99.0	4.26	180	0.096	1.80	5,500	20
1,2-Dichloroethane	99.0	0.98	610		1.53	8,690	20
1,1-Dichloroethylene	97.0	34	600	0.084	1.84	210	25
1,2-Dichloroethylene	96.9	606	208	–	0.48	600	20
1,2 Dicholopropane	113.0	2.31	42	–	2.00	2,700	20
1,3-Dichloropropylene	111.0	3.55	380	–	1.98	2,800	25
Ethylbenzene	106.2	6.44	7	0.071	3.15	152	20
Methylene chloride	84.9	2.03	349	–	1.3	16,700	25
Pyrene	202.3	0.005	–	–	4.88	0.16	26
Styrene	104.1	9.7	5.12	0.075	2.95	300	20
1,1,1,2-Tetrachloroethane	167.8	0.381	5	0.077	3.04	200	20
1,1,2,2-Tetrachloroethane	167.8	0.38	–	–	2.39	2,900	20
Tetrachloroethylene (PCE)	165.8	25.9	–	0.077	2.6	150	20
Tetrachloromethane	153.8	230	–	–	2.64	785	20
Toluene	92.1	6.7	22	0.083	2.73	515	20
Tribromoethane	252.8	0.552	5.6	–	2.4	3,200	30
1,1,1-Trichloroethane	133.4	14.4	100	–	2.49	4,400	20
1,1,2-Trichloroethane	133.4	1.17		–	2.47	4,500	20
Trichloroethylene (TCE)	131.4	9.1	60	–	2.38	1,100	25
Trichlorofluoromethane	137.4	58	667	0.083	2.53	1,100	25
Vinyl chloride	62.5	81.9	2,660	0.114	1.38	1.1	25
Xylenes	106.2	5.1	10	0.076	3.0	198	20

Source: LaGrega et al. 2001; U.S. EPA 1990.

of some COCs are shown in Table 2.2. For comparison, the vapor pressure of water at 25°C and 1 atm is 23.8 mm-Hg, while its vapor pressure is 760 mm-Hg (1 atm) at 100°C, its boiling point.

If the liquid inside a container is a mixture of several volatile compounds, then the vapors of each compound will exert pressure on the liquid. Each vapor-phase compound will exert a different pressure based on its vapor pressure and its mole fraction present in the container. For an ideal solution, the vapor-liquid equilibrium follows Raoult's law as:

$$P_A = \left(P^{vap}\right)\left(x_A\right) \tag{2.10}$$

where P_A is the partial pressure of compound A in the vapor phase, P^{vap} is the vapor pressure of compound A as a pure liquid, and x_A is the mole fraction of compound A in the solution.

The partial pressure is the pressure that a compound would exert if all other gases were not present. This is equivalent to the mole fraction of the compound in the gas phase (y_A) multiplied by the total pressure of the gas (P_t).

$$P_A = P_t \times y_A \tag{2.11}$$

Raoult's law holds only for ideal solutions. In dilute aqueous solutions commonly found in environmental applications, Henry's law, which will be discussed in the next section, is more suitable.

Example 2.7: Vapor Concentration in a Void with the Presence of Free Product

Toluene leaked from a UST at a site and entered the vadose zone. Estimate the maximum toluene concentration (in ppmV) in the pore space of the subsurface. The temperature of the subsurface is 20°C.

SOLUTION:

(a) From Table 2.2, the vapor pressure of benzene is 22 mm-Hg at 20°C.
(b) 22 mm-Hg = (22 mm-Hg) ÷ (760 mm-Hg/1 atm) = 0.0289 atm.
(c) The partial pressure of benzene in the pore space is 0.0289 atm ($28,900 \times 10^{-6}$ atm), which is equivalent to 28,900 ppmV.

DISCUSSION:

1. This 28,900 ppmV value is the vapor concentration in equilibrium with the pure toluene solution. The equilibrium can occur in a confined space or a stagnant condition. If the system is not totally confined, the vapor tends to move away from the source and creates a concentration gradient (the vapor concentration decreases with the distance from the liquid). However, in the vicinity of the solution, the vapor concentration would be at or near this equilibrium value.
2. It is useful to know the values of atmospheric pressure in many units: 1 atm = 14.7 psi = 760 mm-Hg = 29.9 in-Hg = 101.3 kPa = 1×10^6 ppmV.

Example 2.8: Vapor Concentration in a Void with the Presence of Free Product

100 kg of an industrial solvent consisting of 50% (by weight) toluene and 50% ethylbenzene leaked from a UST and entered the vadose zone. The solvent is

found to be present in the void in a "free product" form. Estimate the maximum toluene and ethylbenzene concentrations (ppmV) in the voids. The temperature of the subsurface is 20°C.

SOLUTION:

(a) From Table 2.2, the vapor pressure of toluene (C_7H_8, MW = 92) is 22 mm-Hg and that of ethylbenzene (C_8H_{10}, MW = 106 g/mol = 106 kg/kmol) is 7 mm-Hg at 20°C.

(b) Basis: 100 kg of the solvent
   For 50% by weight of toluene, the percentage by mole (i.e., mole fraction)
   = (moles of toluene) ÷ [(moles of toluene) + (moles of ethylbenzene)] × 100
   = (50 kg/92 kg/kmol) ÷ [(50 kg/92 kg/kmol) + (50 kg/106 kg/kmol)] × 100 = 53.5%
   The partial pressure of toluene in the void can be estimated from Eq. 2.8
   = $(P^{vap})(x_A)$
   = (22)(0.535) = 11.78 mm-Hg = 0.0155 atm = 15,500 ppmV
   The partial pressure of ethylbenzene in the pore space can also be estimated from Eq. 2.8
   = (7)(1 − 0.535) = 3.25 mm-Hg = 0.0043 atm = 4,300 ppmV

DISCUSSION:

The vapor concentrations are those in equilibrium with the solvent. The equilibrium can occur in a confined space or a stagnant phase. If the system is not totally confined, the vapor tends to move away from the source and creates a concentration gradient (the vapor concentration decreases with the distance from the solvent). However, in the vicinity of the solvent, the vapor concentration would be at or near the equilibrium value.

2.5.4   LIQUID-VAPOR EQUILIBRIUM

The free product, as well as its vapor in the void space of the vadose zone, may enter the soil moisture via dissolution or absorption. Equilibrium conditions exist when the rate of the compound entering the soil moisture equals the rate of compound volatilizing from the soil moisture. The COCs in the soil moisture may get adsorbed onto the soil grains through adsorption. Partition coefficients are coefficients that describe how a chemical is distributed (partitioned) between two media. They vary according to the chemical, the media, and the system parameters (i.e., temperature and pressure).

Henry's Law Constant

Henry's law is a concept similar to that of vapor pressure but applied to volatile chemicals dissolved in water (vapor pressure applies to the pure, undissolved chemical). Henry's law is used to describe the equilibrium relationship between the liquid concentration and the corresponding vapor concentration. Henry's law says that under equilibrium conditions, the partial pressure of a volatile chemical in the gas above a liquid is proportional to the concentration of the chemical in the liquid.

$$H_A = \frac{P_A}{C_A} \tag{2.12}$$

where H_A is the Henry's law constant for chemical A, P_A = the partial pressure for chemical A, and C_A is the concentration in water of chemical A.

Since Henry's law constant is a constant number (at a constant temperature and pressure), this equation demonstrates the linear relationship between the concentrations of a chemical in liquid and vapor. Henry's law constant expresses the amount of chemical partitioning between air and liquid at equilibrium. Generally, the greater the Henry's law constant, the more volatile the compound is and thus the more readily it can be removed from a solution. Henry's law constant can be used to assess how a chemical is distributed in the subsurface. For example, if we know the concentration of a chemical in the soil moisture, then we can use Henry's law constant to estimate its vapor concentration in the soil pores.

Henry's law constant can be expressed in many different units. Common units are atm, atm·m³/mole, atm/M, and even dimensionless. Engineers from different backgrounds have different preferences on which units to use. Therefore, the following paragraphs discuss these different units. For example, Henry's law constant can also be expressed as:

$$H_A = \frac{G_A}{C_A} \tag{2.13}$$

where C_A is the chemical A concentration in water, and G_A is the chemical concentration in the gas phase.

Another way of expressing Henry's law constant is

$$H_A = \frac{P_A^{vap}}{Sol_A} \tag{2.14}$$

where P_A^{vap} is the vapor pressure of chemical A, and Sol_A is the solubility of the same chemical at the same temperature.

When the units in the numerator are the same as in the denominator, Henry's law constant is dimensionless. The dimensionless Henry's law constant is often denoted as H*. This dimensionless constant can also be found in the following way:

$$H_A^* = \frac{H_A}{RT} \tag{2.15}$$

where H_A is Henry's law constant for chemical A in units of atm/M (or atm L/mole), R is the universal gas constant, and T is the temperature. Care should be taken to ensure that the units on the right side of the equation cancel out.

Given all the different ways in which Henry's law constant can be expressed, it can be reported in many different units. Table 2.3 is a conversion table for Henry's constant. Note that the dimensionless Henry's constant is not a true dimensionless unit. The actual meaning of Henry's constant in a dimensionless format is (concentration in the vapor phase)/(concentration in the liquid phase), which can be [(mg/L)/(mg/L)].

$$H_A^* = \frac{G_A}{C_A} \tag{2.16}$$

Example 2.9: Henry's Constant and Its Unit Conversions

Estimate Henry's law constant of benzene, in units of atm/M, from its vapor pressure and solubility as shown in Table 2.2. Then convert it to dimensionless and to units of atm.

SOLUTION:

(a) From Table 2.2, the vapor pressure of benzene is 95.2 mm-Hg and the solubility is 1,780 mg/L at 25°C. The molecular weight is 78.1 g/mol. From Eq. 2.14, Henry's constant = vapor pressure ÷ solubility. Converting to appropriate units:

Vapor pressure = (95.2 mm-Hg) × (1 atm/760 mm-Hg) = 0.125 atm
Solubility = (1,780 mg/L) × (1 mole/78,100 mg) =
2.28×10^{-2} mole/L = 2.28×10^{-2}M

Henry's constant = vapor pressure ÷ solubility
= (0.125 atm) ÷ (2.28×10^{-2} M) = 5.48 atm/M

(b) From Table 2.3,
H = H*RT = 5.48 = H*(0.082)(273 + 25) → H* = 0.224 (dimensionless)

(c) Also from Table 2.3,
H = H*RTρ/MW = (0.224)(0.082)(273 + 25)(1,000)/(18) = 304 atm

---

TABLE 2.3
Henry's Constant Conversion Table

Desired unit for Henry's constant	Conversion equation
atm/M, or atm·L/mole	H = H*RT
atm·m³/mole	H = H*RT/1,000
M/atm	H = 1/(H*RT)
atm/(mole fraction in liquid), or atm	H = H*RTρ/MW

Notes: H* = Henry's constant in the dimensionless form
ρ = density of water (for a dilute aqueous solution, 1,000 g/L or 1,000 kg/m³)
MW = equivalent molecular weight of solution (18 g/mol for dilute aqueous solution)
R = 0.08206 atm L/mol/K
T = system temperature in degrees Kelvin
P = system pressure in atm (usually = 1 atm)
M = solution molarity in (g-mol/L)

Source: Kuo and Cordery 1988.

DISCUSSION:

1. The calculated value, 5.48 atm/M, is essentially the same as the value in Table 2.2, i.e., 5.55 atm/M.
2. Values of vapor pressure, solubility, and Henry's constant mentioned in a technical article might come from different sources. Therefore, Henry's constant derived from the ratio of vapor pressure and solubility might not match well with the stated value of Henry's constant.
3. Benzene is a VOC of concern and is shown in most, if not all, databases of Henry's constant values. It may not be a bad idea to memorize that benzene has a dimensionless Henry's constant of ~0.23 under ambient conditions.
4. To convert the Henry's constant of another COC in the database, just multiply the ratio of the Henry's constants (in any units) of that COC and of benzene by 0.23. For example, to find the dimensionless Henry's constant of methylene chloride, first read the Henry's constant for methylene chloride, 2.03 atm/M, and for benzene, 5.55 atm/M, from Table 2.2. Then find the ratio of these two and multiply it by 0.23, as $[(2.03)/(5.55)] \times (0.23) = 0.084$.

Example 2.10: Use Henry's Law to Estimate the Equilibrium Concentrations

The subsurface of a site is impacted by tetrachloroethylene (PCE). A recent soil vapor survey indicates that the soil vapor contained 1,000 ppmV of PCE. Estimate the PCE concentration in the soil moisture. Assume the subsurface temperature is equal to 20°C.

SOLUTION:

(a) From Table 2.2, for PCE, $H = 25.9$ atm/M, and $MW = 165.8$.
Also, the concentration 1,000 ppmV means that the partial pressure
$P_A = 1,000$ ppmV $= 1,000 \times 10^{-6}$ atm $= 1.0 \times 10^{-3}$ atm
Use Eq. 2.12, $P_A = H_A C_A = 1.0 \times 10^{-3}$ atm $= (25.9$ atm/M$) \times (C_A)$
So, $C_A = (1.0 \times 10^{-3}) \div 25.9 = 3.86 \times 10^{-5}$ M
    $= (3.86 \times 10^{-5}$ mole/L$)(165.8$ g/mole$)$
    $= 6.4 \times 10^{-3}$ g/L $= 6.4$ mg/L $= 6.4$ ppm
(b) We can also use the dimensionless Henry's constant to solve this problem.
$H = H^*RT = 25.9 = H^*(0.082)(273 + 20)$
    $H^* = 1.08$ (dimensionless)
Use Eq. 2.5 to convert ppmV to mg/m³:
    G_A (in mg/m³) $= (1,000)(165.8/24.05) = 6,890$ mg/m³ $= 6.89$ mg/L
Use Eq. 2.16, $G_A = H_A^* C_A$
    6.89 mg/L $= (1.08)C_A \rightarrow C_A = 6.4$ mg/L $= 6.4$ ppm

DISCUSSION:

1. A soil vapor survey is a screening tool commonly used in site assessment to define the potential extent of the plume in contaminated soil.
2. Approaches (a) and (b) yield identical results.

3. Henry's constant of PCE is relatively high (five times higher than that of benzene, 1.08 vs. 0.23).
4. A concentration of 6.4 mg/L of PCE in the soil moisture is in equilibrium with a vapor concentration of 1,000 ppmV.
5. The numeric value of the vapor concentration (1,000 ppmV) is much higher than that of the corresponding liquid concentration (6.4 ppm).

Example 2.11: Mass Partition between Vapor and Liquid Phases

A new field technician was sent out to collect a groundwater sample from a monitoring well. He filled only half of the 40-mL sample vial with groundwater impacted by benzene (T = 20°C). The benzene concentration in the collected groundwater was analyzed to be 5 mg/L.
   Determine:

(a) The concentration of benzene in the head space (in ppmV) before the vial was opened.
(b) The percentage of total benzene mass in the aqueous phase of the closed vial.
(c) The true benzene concentration in the groundwater, if a sample free of headspace were collected.

Assume the value of the dimensionless Henry's constant for benzene is equal to 0.22.

SOLUTION:

To simplify this problem, do the calculations on the basis of a 1-L container.

(a) From Henry's law, the concentration of benzene in the head-space = $G = (H)(C)$
   $= (0.22)(5) = 1.1$ mg/L $= 1,100$ mg/m³
   From Eq. 2.5a: G (in ppmV) = G (in mg/m³)/(MW/24.05) = 1,100/(78/24.05) = 340 ppmV
(b) Mass of benzene in the liquid phase = (C)(volume of the liquid)
   $= (5$ mg/L)(1 L × 50%) = 2.5 mg
   Mass of benzene in the headspace = (G)(volume of the air space)
   $= (1.1$ mg/L)(1 L × 50%) = 0.55 mg
   Total mass of benzene = mass in the liquid + mass in the headspace
   $= 2.5 + 0.55 = 3.05$ mg
   Percentage of total benzene mass in the aqueous phase = 2.5/3.05 = 82%
(c) The actual liquid concentration should have been = (total mass of benzene) ÷ (volume of the liquid)
   $= (3.05$ mg)/(0.5 L) = 6.1 mg/L

DISCUSSION:

1. Although the sample volume is only 40 mL, the calculation basis was 1 L to simplify the calculation. The concentration and percent results would have been the same on a 40 mL basis.

2. With the presence of the headspace in the sample bottle, the benzene volatilized from the water, and the apparent liquid concentration was lower than the actual concentration.

2.5.5 SOLID-LIQUID EQUILIBRIUM

Adsorption

Adsorption is the process in which a compound moves from the liquid phase onto the surface of the solid across the interfacial boundary. Adsorption is caused by interactions among three distinct components:

- adsorbent (e.g., vadose zone soil, aquifer matrix, and activated carbon)
- adsorbate (e.g., the COC)
- solvent (e.g., soil moisture and groundwater)

The adsorbate is removed from the solvent and taken by the adsorbent. Adsorption is an important mechanism governing the COC's fate and transport in the environment, as shown in Figure 2.8.

saturated soil

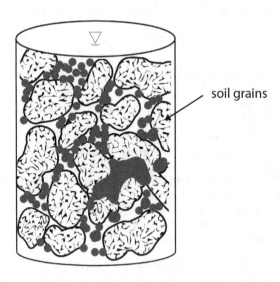

soil grains

● = chemical dissolved in water

❧ = chemical adsorbed to soil

FIGURE 2.8 An illustration of how chemicals can be adsorbed to soil grains as well as dissolved in the soil moisture.

Adsorption Isotherms

For a system where the solid phase and liquid phase coexist, an adsorption isotherm describes the equilibrium relationship between the liquid and solid phases. The "isotherm" indicates that the relationship is for a constant temperature. In practical terms, we use the word "isotherm" to describe an equation.

The most popular isotherms are the Langmuir isotherm and the Freundlich isotherm. Both were derived in the early 1900s. The Langmuir isotherm has a theoretical basis that assumes mono-layer coverage of the adsorbent surface by the adsorbates. The Freundlich isotherm is a semiempirical relationship. For a Langmuir isotherm, the concentration on the soil increases with increasing concentration in the liquid until a maximum concentration on the solid is reached. The Langmuir isotherm can be expressed as follows:

$$S = S_{max} \frac{KC}{1 + KC} \tag{2.17}$$

where S is the adsorbed concentration on the solid surface, C is the dissolved concentration in liquid, K is the equilibrium constant, and S_{max} is the maximum adsorbed concentration. In this book, both symbols "S" and "X" are used for COC concentration in soil. Symbol "S" means "mass of COC/mass of dry soil," while "X" means "mass of COC/mass of soil plus moisture."

The Freundlich isotherm can be expressed in the following form:

$$S = KC^{1/n} \tag{2.18}$$

Both K and 1/n are empirical constants. These constants are different for different adsorbates, adsorbents, and solvents. For a given compound, the values will also be different for different temperatures. When using the isotherms, we should ensure that the units among the parameters and the empirical constants are consistent.

Both isotherms are nonlinear. Incorporating the nonlinear Langmuir or Freundlich isotherm into the mass balance equation to evaluate the COC's fate and transport will make the computer simulation more difficult or more time-consuming. Fortunately, it was found that, in many environmental applications, the linear form of the Freundlich isotherm applies. It is called the linear adsorption isotherm, when 1/n = 1, thus

$$S = K \times C \tag{2.19}$$

which simplifies the mass balance equation in a fate and transport model.

Liquid-Soil Partition Coefficient

For soil-water systems, the linear adsorption isotherm is often written in the following form:

$$S = K_p \times C \quad \text{or} \quad K_p = S/C \tag{2.20}$$

where K_p is the liquid-soil partition coefficient that measures the tendency of a compound to be adsorbed onto the surface of soil or sediment from a liquid phase. It describes how a COC distributes (partitions) itself between the two media (i.e., solid and liquid). Henry's constant, which was discussed earlier, can be viewed as the vapor-liquid partition coefficient.

Organic Carbon Partition Coefficient (K_{oc})

When a chemical adsorbs to the soil, it typically adsorbs to the organic carbon component of soil. The organic carbon partition coefficient (K_{oc}) is the ratio of the concentration of a chemical adsorbed to the organic carbon component of soil to the concentration of the chemical in the aqueous phase. High K_{oc} values indicate a high tendency for a chemical to adsorb to the carbon fraction in the soil. This suggests that the COC has low mobility; it remains adsorbed to the soil rather than moving into a fluid, such as soil moisture. A low K_{oc} indicates a tendency not to adsorb to the carbon fraction in the soil, suggesting high mobility of the COC. For a given organic chemical compound, the partition coefficient is not the same for every soil. The dominant mechanism of organic adsorption is the hydrophobic bonding between the compound and the natural organics associated with the soil. Given that a chemical adsorbs mainly to the organic carbon fraction of the soil, then the K_p is related to the K_{oc} through the fraction of organic carbon (f_{oc}) in soil. When the organic fraction is known, the relationship is

$$K_p = f_{oc} \times K_{oc} \qquad (2.21)$$

The organic carbon partition coefficient (K_{oc}) can be considered as the partition coefficient for the organic compound into a hypothetical pure organic carbon phase. For soil that is not 100% organics, the partition coefficient is discounted by the factor f_{oc}, which is typically in the order of magnitude of 0.001 to 0.02 but can be as high as 0.2. Clayey soil is often associated with more natural organic matter and, thus, has a stronger adsorption potential for organic COCs.

Octanol-Water Partition Coefficient (K_{ow})

K_{oc} is actually a theoretical parameter, and it is the slope of experimentally determined K_p vs. f_{oc} curves. K_{oc} values for many compounds are not readily available. Much research has been conducted to relate them to more commonly available chemical properties such as solubility in water (Sol_w) and the octanol-water partition coefficient (K_{ow}).

The octanol-water partition coefficient is the ratio of the concentration of a chemical dissolved in octanol (an organic solvent) to the concentration of the chemical dissolved in water (i.e., in the aqueous phase). See a listing in Table 2.2. High K_{ow} values indicate that chemicals are likely to adsorb to organic materials and soil rather than be dissolved in water. These chemicals would be considered hydrophobic (from the Greek words for "water" and "fear"). Low K_{ow} chemicals tend to be hydrophilic (from the Greek words for "water" and "love"), stay dissolved in the aqueous phase,

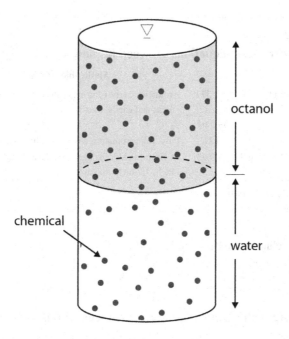

FIGURE 2.9 Illustration of how a chemical (represented by circles) can distribute itself between water and octanol.

and have low adsorption to soil. In soil and groundwater contamination, we would expect a chemical with high K_{ow} to adsorb onto the soil matrix more than dissolving in water. Figure 2.9 shows a container with water and octanol (an LNAPL), plus a chemical dissolved in both. The dots signify the dissolved chemical. With more dots in the octanol column than in the water column, the chemical is shown to have a higher affinity to octanol. Therefore, the chemical has a high K_{ow}.

The octanol-water partition coefficient (K_{ow}) is a dimensionless constant defined by:

$$K_{ow} = \frac{C_{octanol}}{C_{water}} \qquad (2.22)$$

where $C_{octanol}$ is the equilibrium concentration of an organic compound in octanol and C_{water} is the equilibrium concentration of the organic compound in water.

K_{ow} serves as an indicator of how an organic compound will partition between an organic phase and water. Values of K_{ow} range widely, from 10^{-3} to 10^7. There are many correlation equations between K_{oc} and K_{ow} (or solubility in water, Sol_w) reported in the literature. Table 2.4 lists the ones mentioned in an EPA handbook (EPA, 1991). It can be seen that K_{oc} increases linearly with increasing K_{ow} or with decreasing Sol_w on a log-log plot. (Note: Values of K_{ow} for some commonly encountered COCs are provided in Table 2.2.) Organic compounds with low K_{ow} values are hydrophilic (like to stay in water) and have low soil adsorption.

TABLE 2.4

Some Correlation Equations between K_{oc} and K_{ow} (or Sol_w)

Equation	Applicable COCs
$\log K_{oc} = 0.544 (\log K_{ow}) + 1.377$ or	Aromatics, carboxylic acids and esters, insecticides, ureas and uacils, triazines, miscellaneous
$\log K_{oc} = -0.55 (\log Sol_w) + 3.64$	
$\log K_{oc} = 1.00 (\log K_{ow}) - 0.21$	Polycyclic aromatics, chlorinated hydrocarbons
$\log K_{oc} = -0.56 (\log Sol_w) + 0.93$	PCBs, pesticides, halogenated ethanes and propanes, PCE, 1,2-dichlorobenzene

Source: U.S. EPA 1991.

A common relationship between K_{oc} and K_{ow} is

$$K_{oc} = 0.63 \, K_{ow} \tag{2.23}$$

Example 2.12: Solid-Liquid Equilibrium Concentrations

The aquifer underneath a site is impacted by tetrachloroethylene (PCE). A groundwater sample contains 200 ppb of PCE. Estimate the PCE concentration adsorbed onto the aquifer material, which contains 1% of organic carbon. Assume the adsorption isotherm follows a linear model.

SOLUTION:

(a) From Table 2.2, for PCE
$\log K_{ow} = 2.6 \rightarrow K_{ow} = 398$ L/kg
(b) From Table 2.4, for PCE (a chlorinated hydrocarbon)
$\log K_{oc} = 1.00(\log K_{ow}) - 0.21 = 2.6 - 0.21 = 2.39$
$K_{oc} = 245$ mL/g = 245 L/kg
Or, from Eq. 2.23
$K_{oc} = 0.63K_{ow} = 0.63(398) = 251$ mL/g = 251 L/kg
(c) Use Eq. 2.21 to find K_p:
$K_p = f_{oc} K_{oc} = (1\%)(251) = 2.51$ mL/g = 2.51 L/kg
(d) Use Eq. 2.20 to find S, the concentration of PCE in soil:
$S = K_p C = (2.51$ L/kg$)(0.2$ mg/L$) = 0.50$ mg/kg

Discussion:

1. Eq. 2.23 ($K_{oc} = 0.63 \, K_{ow}$), which looks very simple, yields an estimate of K_{oc} (251 kg/L) that is comparable to the value (245 L/kg) from using the correlation equation in Table 2.4.
2. Most technical articles do not talk about the units of K_p, K_{oc}, and K_{ow}. Actually, these partitioning coefficients have units of "(volume of solvent)/(mass of adsorbent)," equal to mL/g or L/kg in most, if not all, of the correlation equations.

2.5.6 Solid-Liquid-Vapor Equilibrium

As mentioned earlier, a NAPL may end up in four different phases as it enters a vadose zone. We have just discussed the equilibrium systems of liquid-vapor and soil-liquid. Now we move one step further to discuss the system, including liquid, vapor, and solid (and free product in some of the applications).

The soil moisture in the vadose zone is in contact with both soil grains and air in the void, and the COC in each phase can travel to the other phases. The dissolved concentration in the liquid, for example, is affected by the concentrations in the other phases (i.e., soil, vapor, and free product). If the entire system is in equilibrium, these concentrations are related by the equilibrium equations mentioned earlier. In other words, if the entire system is in equilibrium and the COC concentration of one phase is known, the concentrations at other phases can be estimated using the equilibrium relationships. Although in real applications, the equilibrium condition does not always exist; the estimate from such a condition serves as a good starting point or as the upper or the lower limit of the real values.

Example 2.13: Solid-Liquid-Vapor-Free Product Equilibrium Concentrations

1,1,1-trichloroethane (1,1,1-TCA) as a free product was found in the subsurface at a site. The soil is silty with an organic content of 2%. The subsurface temperature is 20°C. Estimate the maximum concentrations of TCA (a) in the air void, (b) in the soil moisture, and (c) on the soil grains.

SOLUTION:

(a) Since the free product is present, the maximum vapor concentration will be the vapor pressure of the TCA liquid at that temperature.
From Table 2.2, the vapor pressure of TCA is 100 mm-Hg at 20°C.
Converting to atm and ppmV:
100 mm-Hg = (100 mm-Hg) ÷ (760 mm-Hg/atm) = 0.1316 atm.
$G = 0.1316$ atm = 131,600 ppmV
Use Eq. 2.5 to convert ppmV to mg/m^3, (MW = 133.4 from Table 2.2)
G (in mg/m^3) = $(131,600)[(133.4/24.05)]$ mg/m^3
$G = 730,000$ $mg/m^3 = 730$ mg/L

(b) To find the concentration of 1,1,1-TCA dissolved in the soil moisture, use Henry's constant. From Table 2.2, H = 14.4 atm/M
Convert H to the dimensionless Henry's constant, using Table 2.3:
$H = H^*RT = 14.4 = H^*(0.08206)(273 + 20)$
$H^* = 0.60$ (dimensionless)
Use Eq. 2.16 to find the liquid concentration:
$G = HC = 730$ mg/L = $(0.60)C$
So, $C = 1,220$ mg/L = 1,220 ppm

(c) From Table 2.2 for TCA, log $K_{ow} = 2.49 \rightarrow K_{ow} = 309$ L/kg
From Eq. 2.23,
$K_{oc} = 0.63K_{ow} = 0.63(309) = 195$ mL/g = 195 L/kg
Use Eq. 2.21 to find K_p:

$K_p = f_{oc} K_{oc}$
$= (2\%)(195) = 3.90 \ mL/g = 3.90 \ L/kg$
Use Eq. 2.20 to find the soil concentration, S:
$S = K_p C = (3.90 \ L/kg)(1,220 \ mg/L) = 4,760 \ mg/kg$

DISCUSSION:

Most technical articles do not talk about the units of K_p, K_{oc}, and K_{ow}. Actually, these partitioning coefficients have units of "(volume of solvent)/(mass of adsorbent)," equal to mL/g or L/kg in most, if not all, of the correlation equations.

Example 2.14: Solid–Liquid-Vapor-Free Product Equilibrium Concentrations (Absence of Free Product)

For a subsurface impacted by 1,1,1-trichloroethane (1,1,1-TCA), the soil vapor concentration at a location was found to be 1,316 ppmV. The soil is silty with an organic content of 2%. The subsurface temperature is 20°C. Estimate the maximum concentrations of TCA (a) in the soil moisture and (b) on the soil grains.

SOLUTION:

(a) With a concentration of 1,316 ppmV, this is 100 times smaller than that in Example 2.13:
$G = 1,316 \ ppmV = 7,300 \ mg/m^3 = 7.30 \ mg/L$
With the dimensionless Henry's constant of 0.60 (from Example 2.13):
$G = HC = 7.32 \ mg/L = (0.60)C$
So, $C = 12.2 \ mg/L = 12.2 \ ppm$
(b) With $K_p = 3.90 \ L/kg$ (from Example 2.13),
$S = K_p C = (3.90 \ L/kg)(12.2 \ mg/L) = 47.6 \ mg/kg$

DISCUSSION:

1. The equilibrium relationships ($G = HC$ and $S = K_p C$) are linear. With a vapor concentration of 1,316 ppmV, which is 100 times smaller than that (131,600 ppmV) in Example 2.13, the corresponding liquid and solid concentrations are correspondingly smaller by 100 times. It should be noted that this is only valid when two systems have the same characteristics (i.e., same H and K_p values).
2. The concentrations are based on the assumption that the system is in equilibrium; the actual values would be different if the system were not in equilibrium.

2.6   BIODEGRADABILITY OF COCS

The term biodegradation refers to the transformation of organic compounds by biochemical reactions in the environment. Biodegradation processes can occur aerobically (in the presence of oxygen) or anaerobically (absence of oxygen), depending on the chemical's molecular structure and the environmental conditions. It's important to know these factors to understand whether the chemical compound is degradable or not in the environment. As such, compounds degradable in laboratory tests may not be degradable in the environment. Thus, it is important to know the conditions in the field (temperature and presence/absence of microbes, nutrients, O_2, moisture, and toxins). Furthermore, regarding specific chlorinated solvents, such as PCE and TCE, their degradation pathways are also equally important. For example, PCE, a carcinogen, under the right anaerobic condition, will undergo reductive dechlorination catalyzed by anaerobic bacteria to yield vinyl chloride, a human carcinogen even more potent than PCE.

Bioremediation is a common soil and groundwater remedial alternative that will be discussed in Chapters 5 and 6. Conducting soil sampling and performing treatability tests will help determine if the environment is able to support aerobic and or anaerobic degradation of a specific compound as well as how one can enhance the conditions in the environment to promote biodegradation. Such data can be used in remedial technology design and pilot testing. More on this topic will be discussed in later chapters.

2.7   SOIL MATRICES

So far we have explored a lot of characteristics of COCs in subsurface contamination. A subsurface (i.e., soil and underlying aquifers) is basically a soil matrix built of soil grains. It is a porous medium, typically having connected pores through which fluids can flow. The soil matrices would have significant impacts on the fate and transport of COCs in the subsurface and on the effectiveness of remediation at the site. The important parameters of soil matrices include characteristics of soil grains (e.g., size, type, surface charges), porosity, hydraulic conductivity, and moisture content. Many books on geology/hydrogeology and civil/geotechnical engineering can complement this discussion in more depth. In this chapter, we limit the discussion to the basic concepts necessary for impacted sites, and these concepts will then be used in the coming chapters.

2.7.1   SOIL CLASSIFICATION

There are several soil classification systems. The simplest one is just based on grain size. One of the systems defines clay as having sizes ≤ 0.002 mm (or 2 micrometers (μm), 2 microns); silt between 0.002 and 0.063 mm; sand between 0.063 and 2 mm; and gravel between 2 and 63 mm.

The U.S. Department of Agriculture (USDA) soil textural classification system is based on particle size and the percent distributions of sand, silt, and clay. Figure 2.10 illustrates a USDA soil textural triangle and is used to classify soils by this system.

The unified soil classification system (USCS, not to be confused with the "U.S. customary system of units") is another commonly used one. The classification is based on grain size and plasticity characteristics of soil. The system first divides soil into coarse-grained and fine-grained sizes. If > 50% of the soil is retained on a No. 200 sieve (0.075 mm), it is designated as coarse-grained; on the other hand, it is termed fine-grained if >50% passes a No. 200 sieve. The fine-gained soil is further

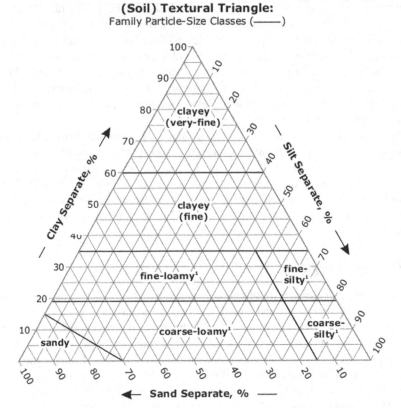

(Soil) Textural Triangle:
Family Particle-Size Classes (———)

¹ Very fine sand fraction (0.05 - 0.1 mm) is treated as silt for Soil Taxonomy family
groupings; coarse fragments are considered the equivalent of coarse sand in the
boundary between silty and loamy classes.

FIGURE 2.10   USDA soil textural triangle. (Source: Schoeneberger et al. 2012.)

divided into different types of silt and clay using its liquid limit and plasticity index
and a Casagrande's Plasticity chart. Organic soil is also included in the fine-grained
soil group.

Fine-grained soil, especially clay, is of greater concern in site remediation. Due to
its smaller particle size, it has a large specific surface area, which is the total surface
area of the grains per unit mass. Clay minerals are aluminosilicates. Surfaces of clay
particles generally have negative charges, so they attract and hold positively charged
species such as metals and soil organic matter. Consequently, clayey soils would
adsorb more soil organics and, consequently, COCs, when compared to sandy soils.

2.7.2   POROSITY AND VOID RATIO

Porosity, ϕ, is the fraction of the total volume of a porous medium (V_t) that is
composed by voids/pores (V_v). Those voids can be the spaces available for water/

groundwater, air, and COCs (in its vapor, dissolved, or NAPL forms) to stay and travel through. Mathematically, porosity is calculated as the ratio of the volume of voids to the total volume of soil.

$$\phi = \frac{V_v}{V_t} \tag{2.24}$$

If the pores are not interconnected, then they are not available for substances to flow through. Therefore, a more useful parameter is the effective porosity, which takes into account the interconnectivity of pores and even the porosity of the soil grains themselves. Typical porosity values are generally expressed as a fraction or a percentage; they may be 0.3, 0.35, 0.4, 0.45, or 30%, 35%, 40%, 45%.

In civil engineering practices, the void ratio is often used. Void ratio is different from porosity. Void ratio (e) is the volume of the voids (V_v) divided by the volume of the solid (V_s), and it is generally expressed as a fraction.

$$e = \frac{V_v}{V_s} \tag{2.25}$$

The porosity cannot be greater than 1, while the void ratio can be greater than 1. Again, the void ratio is the ratio of the void and the volume of the solid portion. Eq. 2.26 relates porosity and the void ratio.

$$e = \frac{\phi}{(1-\phi)}; \qquad \phi = \frac{e}{(1+e)} \tag{2.26}$$

2.7.3 DEGREE OF SATURATION AND WATER/MOISTURE CONTENT

The degree of saturation (S_w) of a soil matrix is defined as the ratio of the volume of water (V_w) to the volume of the void (V_v). It is generally expressed as a percentage and is also known as percent saturation. The degree of saturation for totally dry soil is 0%, while that for soil with all the voids filled with water is 100% (saturated), as for the case of an aquifer.

$$S_w = \frac{V_w}{V_v} \times 100\% \tag{2.27}$$

Volumetric water content or moisture content (ϕ_w) is defined as the ratio of the volume of water (V_w) to the total volume of a porous medium (V_t). In civil engineering practice, gravimetric water content or moisture content of a soil matrix (u) is often used, which is defined as the ratio of the mass/weight of water (M_w) to the mass/weight of the dry soil (M_s). It is generally expressed as a percentage.

$$u = \frac{M_w}{M_s} \times 100\% \tag{2.28}$$

2.7.4 BULK DENSITY

Earlier, we talked about total bulk density (ρ_t), dry bulk density (ρ_b), and specific gravity. The SG value of sand particles ranges around 2.65 and that of inorganic clay around 2.75. The relationship among ρ_t, ρ_b, and u is

$$\rho_t = \rho_b \times (1+u) \qquad\qquad (2.29)$$

Example 2.15: Physical Parameters of a Soil Matrix

The weight of a chunk of moist soil sample is 100.0 lb. The volume of the soil chunk measured before drying is 0.88 ft³. After the sample is dried out in an oven, its weight is 82.0 lb. The specific gravity of the solid is 2.65. Determine (a) water content, (b) total bulk density, (c) dry bulk density, (d) void ratio, (e) porosity, and (f) degree of saturation.

SOLUTION:

(a) Mass of water = (mass of moist soil) − (mass of dry soil) = 100.0 − 82.0 = 18.0 lb
From Eq. 2.28,

$$u = \frac{M_w}{M_s} \times 100\% = \frac{18.0}{82.0} \times 100\% = 22.0\%$$

(b) Total bulk density = ρ_t = (total mass)/(total volume) = 100.0/0.88 = 113.6 lb/ft³

(c) Dry bulk density = ρ_b = (mass of dry soil)/(total volume) = 82/0.88 = 93.2 lb/ft³

(d) Volume of soil grain = (mass of dry soil)/(density of soil grain)
  = 82.0 lb/(2.65 × 62.4 lb/ft³) = 0.496 ft³
Volume of void = (total volume of soil) − (volume of the soil grain) = (0.88 − 0.496) = 0.384 ft³
Void ratio = (volume of the void)/(volume of the soil grain) = 0.384/0.496 = 0.774 = 77.4%

(e) Porosity = ϕ = (volume of the void)/(total volume of soil) = 0.384/(0.88) = 0.436 = 44%

(f) Volume of water = (mass of water)/(density of water) = (18.0 lb)/(62.4 lb/ft3) = 0.288 ft³
Degree of saturation = S_w = (volume of water)/(volume of the void) = (0.288)/(0.384) = 75%

DISCUSSION:

1. The dry bulk density (part (c)) can also be found by using Eq. 2.29 and the values of the total bulk density and water content.
2. The density of the soil grain in part (d) was found as the product of the specific gravity of the solid (2.65) and the density of water (62.4 lb/ft³).
3. Porosity (part (e)) can also be found by using the void ratio and Eq. 2.26.

Example 2.16: Physical Parameters of a Soil Matrix (SI Units)

The average specific gravity of soil grains at a site is 2.65. The porosity is equal to 0.40, and the gravimetric water content is 0.12. On a 1 m³ volume of soil, determine the (a) dry bulk density, (b) total bulk density, (c) volumetric water content, and (d) degree of saturation of the soil.

SOLUTION:

Basis: 1 m³ of soil

(a) Volume of the soil grain = (volume of the soil) × (1 − porosity) = (1.0)(1 − 40%) = 0.6 m³
Mass of the soil grain = (volume)(density) = (0.6 m³)(2.65 × 1,000 kg/m³) = 1,590 kg
Dry bulk density = (mass of dry soil)/(total volume) = 1,590/1.0 = 1,590 kg/m³
(b) Total bulk density = (dry bulk density) × (1 + gravimetric water content) = (1,590)(1 + 0.12) = 1,419 kg/m³
(c) Mass of water = (mass of dry soil) × (gravimetric water content) = (1,590) (0.12) = 190.8 kg
Volume of water = (mass of water) ÷ (density of water) = (190.8)/ (1,000) = 0.191 m³
Volumetric water content = (volume of water)/(volume of the soil) = (0.191)/ (1.0) = 19.1%
(d) Degree of saturation = (volume of water)/(volume of the void) = (0.191)/ (0.4) = 0.48 = 48%

DISCUSSION:

1. The density of soil grain in part (a) was found as the product of the specific gravity of the solid (2.65) and the density of water (1,000 kg/m³).
2. Total bulk density (part (b)) can also be found by using the total mass of soil, which is the sum of the mass of water and mass of soil grain, and the total volume of soil.

2.8 PARTITION OF COCS IN DIFFERENT PHASES

The total mass of COCs in the vadose zone is the sum of the mass in four phases (vapor, moisture, solid, and free product). Let us consider a COC plume in the vadose zone with a volume, V.
From Eq. 2.2,

$$\text{mass of COC dissolved in the soil moisture } = (V_l)(C) = \left[V(\phi_w)\right]C \qquad (2.30)$$

From Eq. 2.4,

$$\text{mass of COC adsorbed onto the soil grains} = (M_s)(S) = \left[V(\rho_b)\right]S \qquad (2.31)$$

From Eq. 2.9,

$$\text{mass of COC in the void space} = (V_a)(G) = \big[(V)(\phi_a)\big]G \qquad (2.32)$$

where V_l is the volume occupied by the moisture, V_a is the volume occupied by air, ϕ_w is the volumetric water content, and ϕ_a is the air porosity. (Note: total porosity, $\phi = \phi_w + \phi_a$.) The total mass of COC (M_t) present in the plume is the sum of the mass in the abovementioned three phases and free product if any. Thus,

$$M_t = V(\phi_w)C + V(\rho_b)S + V(\phi_a)G + \text{mass of the free product} \qquad (2.33)$$

The mass of the free product is simply the volume of the free product multiplied by its mass density. If no free product is present, Eq. 2.33 can be simplified to:

$$M_t = V(\phi_w)C + V(\rho_b)S + V(\phi_a)G \qquad (2.34)$$

To use the equations in this subsection, the following units are suggested: V (in liters), G (mg/L), C (mg/L), S (mg/kg), M_t (mg), ρ_b (kg/L), K_p (L/kg), and η, ϕ_w, ϕ_a, and H (dimensionless).

As mentioned earlier, both "S" and "X" are used for COC concentrations in soil in this book. S is the adsorbed concentration on the solid surface, and X is the COC concentration of a soil sample obtained from laboratory analyses. The symbol "S" means "mass of COC/mass of dry soil," while "X" means "mass of COC/mass of soil plus moisture." When the value of S is not available, the value of X can be used, as the value of S is acceptable in most practical applications (see Example 2.38 in Kuo 2014).

Example 2.17: Mass Partition between Solid and Liquid Phases in an Aquifer

The aquifer underneath a site is impacted by tetrachloroethylene (PCE). The aquifer porosity is 0.4, and the (dry) bulk density of the aquifer material is 1.6 g/cm³. A groundwater sample contains 200 ppb of PCE.
  Assuming that the adsorption follows a linear model, estimate:

(a) The PCE concentration adsorbed on the aquifer material, which contains 1% by weight of organic carbon.
(b) The partition of PCE in two phases, i.e., dissolved phase and adsorbed onto the solid phase.

SOLUTION:

(a) The PCE concentration adsorbed onto the solid has been determined in Example 2.12 as 0.50 mg/kg.
(b) Basis: 1 L aquifer formation

  Mass of PCE in the liquid phase $= (C)[(V)(\phi)] = (0.2)[(1)(0.4)] = 0.08$ mg
  Mass of PCE adsorbed on the solid $= (S)[(V)(\rho_b)] = (0.5)[(1)(1.6)] = 0.8$ mg

Total mass of PCE = mass in the liquid + mass on the solid = 0.08 + 0.8 = 0.88 mg
Percentage of total PCE mass in the aqueous phase = 0.08/0.88 = 9.1%

Discussion: Most of the PCE, 90.9%, in the impacted aquifer is adsorbed onto the aquifer materials. This partially explains why the cleanup of an aquifer takes a long time using the pump-and-treat method.

Example 2.18: Mass Partition between Vapor, Liquid, and Solid Phases

The vapor concentrations of benzene and pyrene in the void space of the vadose zone underneath a landfill are 100 ppmV and 10 ppbV, respectively. The total porosity of the vadose zone is 40%, and 30% of the voids are occupied by water. The (dry) bulk density of the soil is 1.6 g/cm³ and the total bulk density is 1.8 g/cm³. Assuming no free product is present, determine the mass fractions of each COC in the three phases (i.e., void, moisture, and solid phases). The values of the dimensionless Henry's constant for benzene and pyrene are 0.22 and 0.0002, respectively. The values of K_p for benzene and pyrene are 1.28 and 717, respectively.

SOLUTION:

Basis: 1 m³ of soil

	Benzene	Pyrene
(a) Determine the mass in the voids		
MW (g/mol)	78	202
G (ppmV)	100	0.01
G (mg/m³)	324.3	0.084
Air void (m³) = 0.40*(1 − 0.3)	0.28	0.28
Mass in void (mg)	90.8	0.024
(b) Determine the mass dissolved in the liquid		
H (dimensionless)	0.22	0.0002
C (mg/m³) = G/H	1,473	420
Liq. volume (m³) = 0.40×0.3	0.12	0.12
Mass in liquid (mg)	176.8	50.4
(c) Determine the mass adsorbed onto the solid		
K_p (L/kg)	1.28	717
C (mg/L)	1.47	0.42
S (mg/kg) = K_p×C (mg/L)	1.88	301
Soil mass (kg) = (1 m³)(ρ_b)	1,600	1,600
Mass on solid surface (mg)	3,008	4.82×10^5
(d) Determine the total mass in three phases		
Total COC (mg)	3,276	4.82×10^5
(e) Determine the mass fraction in each phase		
% in void	2.8	5×10^{-6}
% in moisture	5.4	0.01
% in solid	91.8	99.9

DISCUSSION:
1. A spreadsheet is a good way to solve a problem such as this one with repetitive calculations.
2. For both compounds, most of the COCs are attached to the solid (91.8% for benzene and 99.9% for pyrene). This is especially true for pyrene that has very high K_p and low H values. The vapor concentration of pyrene is extremely low, while its concentration in soil is very high.

2.9   SUMMARY

This chapter discussed the characteristics of chemicals and soils that need to be known for subsurface site investigation and remediation. The major principles covered were:

- Sources of contamination
- Types of common chemical contaminants
- Units of concentration and mass
- Physical and chemical properties of chemicals
- Types of soils and soil characteristics
- Partitioning of chemicals in different phases

The following chapters will draw upon the principles covered in this chapter and apply them to activities performed by scientists and engineers at contaminated sites.

2.10   PROBLEMS AND ACTIVITIES

2.1.   For the following chemicals, conduct an internet search and describe (i) uses of the chemical, (ii) physical characteristics, (iii) exposure routes, (iv) health symptoms and target organs. Possible sources are the National Institute for Occupational Safety and Health (NIOSH) Pocket Guide and the U.S. EPA.
(a) PCE, (b) TCE, (c) Benzene, (d) Toluene, and (e) Hexavalent Chromium (Cr^{6+})

2.2.   Convert the following units of contaminant concentrations in water to part per million (ppm):
a) 2.0 mg/L of TPH
b) 385 µg/L of ethylbenzene
c) 29 parts per billion (ppb) of benzene

2.3.   Convert the following units of contaminant concentration in water to micrograms per liter (µg/L):
a) 87 parts per billion (ppb) of benzene
b) 0.35 mg/L of toluene
c) 4.58 ppm of chromium

2.4.   Calculate the volume that 1 mole of an ideal gas occupies at a temperature of 40°C and a pressure of 0.97 atm (note that since the pressure is less than atmospheric pressure, 1 atm, the pressure is considered to be a vacuum).

2.5. Based on your answer to problem 2.4, convert a concentration of PCE of 48 μg/m^3 to ppmV.

2.6. To compare how temperature and pressure affect a measurement of a gas concentration, calculate 300 ppmV of methyl ethyl ketone in units of mg/m^3 in (a) the atmosphere at 20°C and 1 atm and (b) flowing inside a pipe at 50°C and 1.2 atm.

2.7. A bottle containing 450 mL of the liquid chemical methylene chloride (CH_2Cl_2, specific gravity 1.33) is left open in a room with dimensions 5 m × 5 m × 3.6 m. A person returning to the room found that only 350 mL of the chemical was left in the bottle, so 100 mL evaporated. (a) If all of the volatilized methylene chloride remained in the room (the worst-case scenario), what would be the concentration of methylene chloride in the air inside the room, in units of mg/m^3? (b) If inside the room the temperature is 25°C and the pressure is 1 atm, what is the volume occupied by 1 mole of methylene chloride, and what is the concentration of methylene chloride in units of ppmV? (c) How does this concentration compare to the Occupational Safety and Health Administration's (OSHA) 8-hour time-weighted average (TWA) permissible exposure limit (PEL) of 25 ppmV?

2.8. A young boy went into a playground and played with dirt impacted with ethyl benzene. During his stay at the site, he ingested a mouthful (about 1.0 cm^3) of soil containing 6.0 ppm ethyl benzene. The bulk density of the soil is 1.8 g/cm^3. How much ethyl benzene did he ingest?

2.9. On a daily basis, an adult drinks water containing 10 ppb benzene and inhales air containing 10 ppbV benzene (at 20°C and 1 atm). Which system (ingestion or inhalation) is exposed to more benzene? The typical water intake rate for an adult is 2.0 L/day, and the air inhalation rate is 15.2 m^3/day.

2.10. Convert the Henry's constants in Table 2.2 for TCE and PCE to dimensionless Henry's constants.

2.11. A 200.0 L drum contains 150 L of water with dissolved styrene. The concentration of styrene in the headspace (gas phase) above the water was measured to be 0.00040 atm (this is a partial pressure). Refer to the Henry's constant and molecular weight of styrene in Table 2.2. What would be the concentration of styrene in the water in units of mg/L?

2.12. Soil vapor sampling was conducted in the city of Spring Valley, resulting in 0.42 μg/m^3 of TCE and 0.79 μg/m^3 of PCE. With the dimensionless Henry's constants found in problem 2.10, calculate the estimated concentrations of TCE and PCE in groundwater directly beneath the soil gas samples, in units of μg/L.

2.13. A chemical is placed in a beaker containing 15 g of soil and 500 mL of water. At equilibrium, the chemical is found in the soil at a concentration of 130 mg/kg. The equilibrium concentration of this same chemical in the water is 310 μg/L. What is the soil-water partition coefficient for this chemical in this soil?

2.14. A chemical has a soil-water partition coefficient of 9,800 L/kg, or (mg/kg)/(mg/L). If the concentration of this chemical in water is found to be 67 μg/L, at equilibrium, what is the concentration adsorbed to the soil?

2.15. Classify a soil sample (25% clay, 35% sand, and 40% silt) using the USDA soil texture triangle.

2.16. The aquifer underneath a site is impacted by the chemical styrene. The aquifer porosity is 0.20, and the dry bulk density of the aquifer material is 1.6 g/cm³. A groundwater sample contains 950 ppb of styrene. (a) Estimate the styrene concentration adsorbed onto the aquifer material, which contains 2.0% of organic carbon. (b) For a 1.0 L aquifer formation, calculate the mass of styrene adsorbed onto the soil. (c) For a 1.0 L aquifer formation, calculate the mass of styrene dissolved in groundwater. (d) Compare the presence of styrene in the two phases and explain what would cause the difference.

REFERENCES

Davis, M. L., and Masten, S. J. (2013). *Principles of Environmental Engineering and Science*. McGraw-Hill Education, New York.

DTSC. (2019). "Candidate Chemical List Frequently Asked Questions." *Department of Toxic Substances Control*, https://dtsc.ca.gov/scp/candidate-chemical-list-frequently-asked-questions/ (Dec. 31, 2019).

Erickson, B. E. (2017). "How Many Chemicals Are in Use Today?" *Chemical & Engineering News (American Chemical Society)*, 95(9), 23–24.

Kuo, J. (2014). *Practical Design Calculations for Groundwater and Soil Remediation*. CRC Press, Boca Raton.

Kuo, J. F., and Cordery, S. A. (1988). "Discussion of Monograph for air Stripping of VOC from Water." *ASCE Journal Environmental Engineering*, 114(5), 1248–1250.

LaGrega, M. D., Buckingham, P. L., and Evans, J. C. (2001). *Hazardous Waste Management*. McGraw-Hill, New York.

MDEQ. (2020). "Underground Storage Tanks." *Mississippi Department of Environmental Quality*, https://www.mdeq.ms.gov/water/groundwater-assessment-and-remediation/underground-storage-tanks/ (Jan. 27, 2020).

Schoeneberger, P. J., Wysocki, D. A., Benham, E. C., and Soil Survey Staff. (2012). *Field Book for Describing and Sampling Soils*, Version 3.0. Natural Resources Conservation Service, National Soil Survey Center, Lincoln, NE.

U.S. EPA. (1990). *CERCLA Site Discharges to POTWs Treatability Manual*. EPA/540/2-90-007. U.S. Environmental Protection Agency, Washington, DC.

U.S. EPA. (1991). *Site Characterization for Subsurface Remediation*. EPA/625/4-91/026. U.S. Environmental Protection Agency Office of Research and Development, Washington, DC.

U.S. EPA. (2016a). *Operating and Maintaining Underground Storage Tank Systems*. EPA 510-K-16-001. U.S. Environmental Protection Agency, Washington, DC.

U.S. EPA. (2016b). *EPA in Illinois, Galena Train Derailment*. https://19january2017snapshot.epa.gov/il/galena-train-derailment_.html (Jun. 22, 2019).

U.S. EPA. (2018). *USS Lead Superfund Site*. https://www.epa.gov/uss-lead-superfund-site/uss-lead-photo-gallery (Jan. 27, 2020).

Yaws, C. L. (2012). *Yaws' Handbook of Properties for Aqueous Systems*. https://app.knovel.com/hotlink/toc/id:kpYHPAS006/yaws-handbook-properties/yaws-handbook-properties.

3 Laws, Regulations, and Risk Assessment Relevant to Site Assessment and Remediation

3.1 REGULATORY FRAMEWORK

The U.S. Congress passes laws that govern the country. For the laws related to the protection of the environment and public health, Congress has authorized the Environmental Protection Agency (EPA) as a regulatory agency to create and enforce regulations to implement these laws. In addition, a number of Presidential Executive Orders (EOs) play a central role in EPA's activities.

A proposed regulation is first published for comments in the Federal Register (FR), the official daily publication for rules, proposed rules, and notices of federal agencies and organizations, as well as EOs and other presidential documents. Once a final decision is issued, the final regulation is codified and incorporated into the U.S. Code of Federal Regulations (CFR). Title 40 (Protection of Environment) of the CFR deals with EPA's mission of protecting human health and the environment. These regulations are mandatory requirements that can apply to individuals, businesses, state or local governments, nonprofit institutions, or others.

A summary of the U.S. environmental laws and EOs can be found at https://www.epa.gov/laws-regulations/laws-and-executive-orders.

The laws and regulations most relevant to site assessment and remediation include the Comprehensive Environmental Response, Compensation, and Liability Act (CERCLA) of 1980 and the Resource Conservation and Recovery Act (RCRA) of 1976.

3.2 CERCLA/SUPERFUND

Contaminated sites usually result from improper management of hazardous waste. In the late 1970s, toxic waste dumps such as Love Canal in Niagara Falls, New York, alerted the public about human health and environmental risks posed by contaminated sites. Many of these sites were abandoned, and little or no effort was made to remediate them. The CERCLA of 1980 is a federal program used to clean up the nation's uncontrolled or abandoned hazardous waste sites as well as accidents, spills, and other emergency releases of pollutants into the environment. Because of the large amount of funding allocated to these hazardous waste sites, CERCLA is nicknamed

Superfund. Superfund is administered by the EPA in cooperation with state and tribal governments. The objectives of the Superfund are to (1) protect human health and the environment by cleaning up contaminated sites; (2) make responsible parties pay for cleanup work; (3) involve communities in the investigation and remediation processes; and (4) return contaminated sites to productive use. The Superfund Amendments and Reauthorization Act (SARA) of 1986 reauthorized CERCLA to continue cleanup activities around the country. Title III of SARA authorized the Emergency Planning and Community Right-to-Know Act (EPCRA).

There are two basic types of responses to manage contaminated sites. The short-term responses are to handle emergency oil spills or chemical releases, while long-term responses would apply remedial actions to handle complex sites. The Superfund cleanup is a multiphase process that includes nine phases. The approach provides guidance that is mostly applicable to all contaminated sites. These nine phases are (U.S. EPA 2011):

    I. Preliminary assessment/site inspection – site assessment
    II. National priorities list (NPL) site listing process
    III. Remedial investigation (RI) /feasibility study (FS) – site characterization
    IV. Record of decision (ROD) – remedy decision
    V. Remedial design (RD)/remedial action (RA)
    VI. Construction completion
    VII. Postconstruction completion – operation and maintenance (O&M)
    VIII. National priorities list deletion
    IX. Site reuse/redevelopment

3.2.1 SITE ASSESSMENT AND NPL LISTING

The activities in the site assessment phase include preliminary assessment (PA) and site investigation (SI). The PA is to evaluate if the site conditions pose human health and environmental risks. This evaluation includes gathering and analyzing historical/existing information to determine if short-term responses are required. An SI, if deemed necessary, is then conducted to test the environmental media (i.e., air, water, and soil) at the site to determine the presence, release, and threat of hazardous substances.

The data/information collected in this PA/SI phase are then used to evaluate the extent of risks posed by the site using EPA's Hazard Ranking System (HRS). Sites with elevated risks are proposed for inclusion on the NPL that includes the most serious sites identified for long-term cleanup. As of April 1, 2020, there are 1,335 NPL sites in the United States.

3.2.2 RI/FS – SITE CHARACTERIZATION

After a site is added to the NPL, a RI/FS is performed. A RI is to collect data to characterize site conditions, determine the nature of the waste, and assess the human health and environmental risks. Treatability testing may be needed to evaluate the potential performance and cost of the treatment technologies being considered. An FS is concurrently conducted for the development, screening, and detailed evaluation of alternative remedial actions.

The FS consists of three major phases: (1) identification and initial screening of remedial technologies, (2) development of remedial alternatives, and (3) detailed analysis of remedial alternatives. During the initial screening phase, the full range of available technologies should be evaluated based on cost, effectiveness, and implementability. A single technology is usually not enough to remediate a site in which more than one medium is impacted (e.g., groundwater and soil). In addition, some remedial technologies may transfer the pollutants from one medium to another (e.g., from groundwater to air by air stripping). Consequently, the screened technologies are combined into remedial alternatives by taking into account the scope, characteristics, and complexity of the site problems being addressed (U.S. EPA 1994). A detailed analysis is then conducted on remedial alternatives.

Based on the results of the FS portion of this phase, the EPA develops a proposed plan for cleaning up the site. The community is notified and can comment on the proposed plan. A public meeting may be held to discuss the proposed plan.

3.2.3   Remedy Decisions

The chosen remedial alternative and the reasons for the selection are set forth in the ROD, a public document. It contains information on (1) history, description, and characteristics of the site, (2) contaminated media and the contaminants present, (3) past, present, and enforcement activities, (4) community participation, (5) description of the response actions to be taken, and (6) the remedy selected for cleanup.

3.2.4   Remedial Design/Remedial Action

During the remedial design phase, the technical specifications for cleanup remedies and technologies, including engineering drawings and specifications, are developed. More sampling may be needed to further define the extent of contamination (e.g., location and amount of contaminants).

Upon the completion of the RD phase, the RA phase starts. The RA phase starts the site cleanup and involves the actual construction and implementation of remedial technologies. The RD/RA is based on the specifications described in the ROD.

3.2.5   Construction Completion and Postconstruction Completion

Construction completion marks a milestone and indicates that all needed physical construction for cleanup at an NPL site is complete, though final cleanup levels may not have been reached.

After the construction completion phase, postconstruction activities are implemented to ensure that cleanup actions protect human health and the environment over the long term. The activities include operation and maintenance (O&M) of remediation systems to ensure that they are in working order, imposing site restrictions (e.g., institutional controls) to minimize the potential for exposure to contamination, and implementing routine monitoring and reviewing of the site data to ensure the effectiveness of the cleanup.

3.2.6 NPL DELETION AND SITE REUSE/REDEVELOPMENT

A contaminated site may be deleted from the NPL once the cleanup goals have been achieved and the site becomes fully protective of human health and the environment.

The ultimate goal of a cleanup program is to return the site to beneficial uses. The site cleanup should be consistent with its likely future use. Consideration of reuse can occur at any point in the cleanup process, from SI activities to NPL deletion. EPA's Superfund Redevelopment Initiative is a nationally coordinated effort to ensure that EPA and the community partners have an effective process and the necessary information and tools to return hazardous waste sites to productive use.

A brownfield is a property in which its expansion, redevelopment, or reuse may be complicated by the presence or potential presence of a hazardous substance/pollutant/contaminant. There are more than 450,000 brownfields in the United States. EPA's Brownfields and Land Revitalization Program (not part of Superfund) was incepted in 1995 to empower states, communities, and other stakeholders to work together to prevent, assess, safely clean up, and sustainably reuse brownfields.

3.3 RCRA

The RCRA, signed into law in 1976, created the framework for hazardous and nonhazardous waste management programs in the United States. RCRA is the primary law of the United States that gives the EPA the authority to control hazardous waste with the "cradle-to-grave" approach. This means the management of hazardous waste from its generation, transportation, treatment, storage to the final disposal. With this, EPA develops regulations, guidance, and policies to ensure safe management and cleanup of solid and hazardous waste, and programs to encourage source reduction and beneficial reuse. Those regulations can be found in Title 40 of the CFR, parts 239–282. RCRA has been amended several times, including the Hazardous and Solid Waste Amendments (HSWA) of 1984.

The goals set by RCRA are (1) to protect human health and the environment from the potential hazards of waste disposal, (2) to conserve energy and natural resources; (3) to reduce the amount of waste generated, and (4) to ensure that wastes are managed in an environmentally sound manner (U.S. EPA 2014).

RCRA states that "solid waste" means any garbage or refuse, sludge (from a wastewater treatment plant, water supply treatment plant, or air pollution control facility) and other discarded material resulting from industrial, commercial, mining, and agricultural operations, and from community activities. It should be noted that solid waste is not limited to wastes that are physically solid; many solid wastes are liquid, semisolid, or containerized gaseous material.

RCRA consists of ten subtitles (A through J). Subtitle D of RCRA is dedicated to nonhazardous solid waste requirements, Subtitle C focuses on hazardous solid waste, and Subtitle I is the regulation of underground storage tanks (USTs).

3.3.1 NONHAZARDOUS WASTE

RCRA Subtitle D focuses on state and local governments as the primary planning, regulating, and implementing entities for the management of nonhazardous solid waste (e.g., household garbage and nonhazardous industrial solid waste). Title 40 of the CFR, parts 239–259, contains the regulations for solid waste. These regulations ban the open dumping of waste and developed federal criteria for proper design and operation of municipal waste and industrial waste landfills. They include design criteria, location restrictions, financial assurance, corrective action (cleanup), and closure requirements.

3.3.2 HAZARDOUS WASTE

The regulations governing hazardous waste identification, classification, generation, management, and disposal are found in Title 40 of the CFR, parts 260–273. These regulations set criteria for hazardous waste generators; transporters; and treatment, storage, and disposal facilities (TSDFs). This includes permitting requirements, enforcement, and corrective action or cleanup. As of 2009, there were approximately 460 TSDFs, 18,000 transporters, and 14,700 large quantity generators (LQGs) on record (U.S. EPA 2014).

Hazardous wastes can be defined as wastes with properties that make them dangerous or potentially harmful to human health or the environment. However, this simple narrative definition would not be sufficient for the development of a regulatory framework capable of ensuring adequate protection. Determining whether a waste is hazardous is paramount because only those wastes that have specific attributes are subject to RCRA Subtitle C regulations. Making such determination is a complex task that is a central component of the hazardous waste management regulations. Readers should refer to relevant RCRA regulations or the *RCRA Orientation Manual 2014* (U.S. EPA 2014) for details. This subsection presents a brief coverage of the EPA's hazardous waste identification process.

For a material to be *classified* as a hazardous waste, it must be a solid waste. However, hazardous wastes can actually be liquids, solids, containerized gases, or sludge, and by-products of manufacturing processes or simply discarded commercial products, like cleaning fluids or pesticides. The second step of the identification process is to determine if the waste is specifically excluded from regulation as a solid or hazardous waste. If it is not excluded, it should be determined whether it is a listed waste or characteristic waste. The last step is to determine if the waste is delisted.

3.3.3 IDENTIFICATION PROCESS OF HAZARDOUS WASTE

RCRA "listed wastes" are wastes from common manufacturing and industrial processes, specific industries, and can be generated from discarded commercial products. They are on one of the four EPA hazardous wastes lists: F-list (nonspecific source wastes) in 40 CFR 261.31; K-list (source-specific wastes) in 40 CFR 261.32; and P-list and U-list (discarded commercial chemical products) in 40 CFR 261.33.

RCRA "characteristic wastes" are those exhibiting at least one of the following four characteristics: ignitability, corrosivity, reactivity, or toxicity. Wastes that are hazardous due to ignitability include liquids with flash points below 60°C (140°F), nonliquids that cause fire through specific conditions, ignitable compressed gases, and oxidizers. D001 is the waste code for ignitable hazardous wastes. There are EPA test methods available for ignitability (i.e., SW-846 Test Methods 1010A, 1020B, and 1030). EPA's SW-846 Compendium (*Test Methods for Evaluating Solid Waste: Physical/Chemical Methods Compendium*) is EPA's official collection of methods for use in complying with RCRA regulations. It consists of more than 200 analytical methods for sampling and analyzing waste and other matrices (e.g., air and water). These methods are divided into sections, or "series," according to the type of method, analyte, and technique used. The entire Compendium is available online (https://www.epa.gov/hw-sw846/sw-846-compendium).

Wastes that are hazardous due to corrosivity include aqueous wastes with a $pH \leq 2$, a $pH \geq 12.5$, or based on the liquid's ability to corrode steel. D002 is the waste code for corrosive hazardous wastes. The SW-846 Test Method 1110A determines corrosivity toward steel.

The reactivity characteristic identifies wastes that readily explode or undergo violent reactions. There are no test methods for reactivity. Instead, EPA uses narrative criteria to define most reactive wastes. The criteria include: (1) it can explode or violently react when exposed to water or under normal handling conditions; (2) it can create toxic fumes or gases at hazardous levels when exposed to water or under normal waste handling conditions; (3) it can explode if heated under confinement or exposed to a strong igniting source, or it meets the criteria for classification as an explosive under Department of Transportation (DOT) rules; and it generates toxic levels of sulfide or cyanide gas when exposed to a pH range of 2–12.5. D003 is the waste code for reactive hazardous wastes.

To evaluate whether a specific waste is likely to leach toxic chemicals into groundwater, a lab procedure known as the Toxicity Characteristic Leaching Procedure (TCLP) (SW-846 Test Method 1311) is conducted. During the TCLP procedure, the waste sample is extracted with an extraction fluid for 48–72 hours. At the completion of the extraction, the extract will be analyzed if it contains any of 40 different toxic chemicals exceeding the specified regulatory levels (see Table 3.1). If the extract contains a concentration exceeding its corresponding limit, the waste exhibits the toxicity characteristic (TC) and carries the waste code associated with that compound or element. It is considered as a "toxicity characteristic" hazardous waste.

Once a hazardous waste is generated, it may become mixed with other wastes, be treated and produce residues, or be spilled. RCRA provides special regulatory provisions to address the regulatory status of hazardous waste mixtures; treatment, storage, and disposal residues; and contaminated media and debris. These provisions are known as the "mixture rule," the "derived-from rule," and the "contained-in policy" (U.S. EPA 2014).

TABLE 3.1
TCLP Regulatory Levels

Waste Code	Contaminant	Concentration (mg/L)
D004	Arsenic	5.0
D005	Barium	100.0
D018	Benzene	0.5
D006	Cadmium	1.0
D019	Carbon tetrachloride	0.5
D020	Chlordane	0.03
D021	Chlorobenzene	100.0
D022	Chloroform	6.0
D007	Chromium	5.0
D023	o-Cresol*	200.0
D024	m-Cresol*	200.0
D025	p-Cresol*	200.0
D026	Total Cresol*	200.0
D016	2,4-D	10.0
D027	1,4-Dichlorobenzene	7.5
D028	1,2-Dichloroethane	0.5
D029	1,1-Dichloroethylene	0.7
D030	2,4-Dichlorotoluene	0.13
D012	Endrin	0.02
D031	Heptachlor (and its epoxide)	0.008
D032	Hexachlorobenzene	0.13
D033	Hexachlorobutadiene	0.5
D034	Hexachloromethane	3.0
D008	Lead	5.0
D013	Lindane	0.4
D009	Mercury	0.2
D014	Methoxychlor	10.0
D035	Methylethylketone	200.0
D036	Nitrobenzene	2.0
D037	Pentachlorophenol	100.0
D038	Pyridine	5.0
D010	Selenium	1.0
D011	Silver	5.0
D039	Tetrachloroethylene	0.7
D015	Toxaphene	0.5
D040	Trichloroethylene	0.5
D041	2,4,5-Trichlorophenol	400.0
D042	2,4,6-Trichlorophenol	2.0
D017	2,4,5-TP (Silver)	1.0

3.3.4 Hazardous Waste Generators

RCRA regulations broadly define a generator as any person, by site, who first creates or produces a hazardous waste or first brings a hazardous waste into the RCRA Subtitle C system. They include various types of facilities and businesses ranging from large manufacturing operations, universities, and hospitals to small businesses, such as dry cleaners and auto body repair shops, and laboratories. Based on the waste generation rates, there are three categories of hazardous waste generators. The LQGs are those facilities that generate $\geq$1,000 kg of hazardous waste or $\geq$1 kg of acutely hazardous waste per calendar month. All other generators that produce >100 kg and <1,000 kg of hazardous waste per calendar month and accumulate on-site <6,000 kg of hazardous waste at any time are small quantity generators (SQGs). The conditionally exempt small quantity generators are those facilities that produce <100 kg hazardous waste or <1 kg of acutely hazardous waste per calendar month.

Although the requirements for SQGs may be less extensive than those for LQGs, all need to comply with good practices of proper management, emergency planning, and personnel training. With regard to the accumulation of waste on-site, an SQG generally may accumulate waste for $\leq$180 days, while $\leq$90 days for an LQG.

3.3.5 Hazardous Waste Transporters, Manifest, and TSDFs

The generated hazardous waste at a generator facility often needs to be shipped to an off-site TSDF for further management. A hazardous waste transporter under Subtitle C is any person engaged in the off-site transportation of hazardous waste within the United States, if such transportation requires a Uniform Hazardous Waste Manifest. Hazardous waste transporters must comply with both EPA and DOT regulations.

The Uniform Hazardous Waste Manifest (EPA Form 8700-22) plays a crucial part in the "cradle-to-grave" hazardous waste management system. The manifest allows all parties involved in hazardous waste management (i.e., generators, transporters, TSDFs, EPA, state agencies) to track the movement of hazardous waste from the generator's site to the site where the waste will be treated, stored, or disposed of. An RCRA manifest contains (1) the required information of the hazardous waste generator, transporter(s), and designated TSDF; (2) DOT description of the waste's hazards; and (3) quantity of the waste transported and the type of container.

TSDFs are the last link in the "cradle-to-grave" hazardous waste management system. The requirements for TSDFs are more extensive than the standards for generators and transporters. They include general facility operating standards as well as technical standards. The technical standards address the diversity of hazardous waste operations being conducted around the country by guiding facilities in the proper design, construction, operation, maintenance, and closure of a variety of hazardous waste treatment, storage, and disposal units (U.S. EPA 2014, 2018).

3.3.6 Underground Storage Tanks

An underground storage tank system is a tank and any underground piping connected to the tank that has at least 10% of its combined volume underground. Approximately 550,000 USTs nationwide store petroleum or hazardous substances.

The greatest potential threat from a leaking UST (LUST) is the contamination of underlying soil and groundwater, the source of drinking water for nearly 50% of the U.S. population. The federal UST regulations apply only to UST systems that store petroleum or certain hazardous substances, and a complete version of the law that governs USTs can be found in the U.S. Code, Title 42, Chapter 82, Subchapter IX. Subtitle I of the RCRA addresses the problems of LUST systems. The minimization of problems related to LUSTs can be achieved through (1) meeting UST regulatory requirements for installation and operation of USTs, (2) prevention and timely detection of releases, and (3) taking timely and effective response/cleanup actions of UST releases and site remediation. More discussion on site assessment/characterization and conceptual site models are presented in Chapter 4.

3.4 APPLICABLE OR RELEVANT AND APPROPRIATE REQUIREMENTS (ARARS)

The EPA has developed the *CERCLA Compliance with Other Environmental Laws Manual* (U.S. EPA 1988, 1989) to provide guidance for planning response actions under CERCLA. The guidance assists in the selection of on-site remedial actions that, at a minimum, attain, waive, or meet the applicable or relevant and appropriate requirements (ARARs) of the RCRA, Clean Water Act (CWA), Safe Drinking Water Act (SDWA), Clean Air Act (CAA), and other Federal and State environmental laws, as required by Section 121 of CERCLA. The establishment of cleanup goals for a specific site is a complex process. In addition to technical aspects for site remediation, one should also consult with regulatory experts and agencies on regulatory requirements. This section presents a very brief summary just to provide readers with some rough idea about ARARs for site remediation.

"Applicable" requirements mean those cleanup standards that specifically address a hazardous substance, pollutant, contaminant, remedial action, location, or other circumstance found at a CERCLA site. These cleanup standards can also be standards of control or other substantive environmental protection requirements, criteria, or limitations promulgated under federal environmental or state environmental or facility siting law. An "applicable requirement" is a requirement that a private party would have to comply with by law if the same action were being undertaken apart from CERCLA authority. If a requirement is not "applicable," it still may be relevant and appropriate. "Relevant and appropriate requirements" mean that those cleanup standards address problems or situations sufficiently similar to those encountered at the CERCLA site that their use is well suited to the particular site. ARARs are identified on a site-by-site basis for all on-site response actions where CERCLA authority is the basis for cleanup. Different ARARs that may apply to a site, and its remedial action, should be identified at multiple points in the remedy selection process.

ARARs are classified into three categories: (1) chemical-specific, (2) location-specific, and (3) action-specific, depending on whether the requirement is triggered by the presence or emission of a chemical, by a vulnerable or protected location, or by a particular action. "Chemical-specific ARARs" are typically health- or risk-based numerical values or methodologies, which, when applied to site-specific conditions, are expressed as numerical values that represent cleanup standards (i.e., the acceptable concentration of a chemical at the site). "Location-specific ARARs" are restrictions on the concentration of hazardous substances (e.g., RCRA land disposal restrictions prohibiting hazardous waste placement into landfills) or the conduct of activities in environmentally sensitive areas (e.g., floodplains, wetlands, and locations where endangered species or historically significant cultural resources are present). "Action-specific ARARs" are usually technology- or activity-based requirements or limitations on actions/conditions taken with respect to specific hazardous substances. Action-specific ARARs do not determine the remedial alternative; instead, they indicate how a selected alternative must be achieved.

3.4.1 CLEANUP STANDARDS

Section 121 of CERCLA requires the selected remedial action be protective of human health and the environment. Determination of protectiveness (i.e., cleanup standards) involves risk assessment, considering both ARARs and to-be-considered materials (TBCs). TBCs are nonpromulgated advisories or guidance issued by the federal or state government. Risk assessment includes consideration of site-specific factors such as types of hazardous substances present, the potential for exposure, and the presence of sensitive populations (more coverage on risk assessment are provided in Section 3.5). Acceptable exposure levels are generally determined by ARARs, if available, and the following factors: (1) for systemic toxicants, concentration levels to which the human population (including sensitive subgroups) could be exposed on a daily basis without appreciable risk of significant adverse effects during a lifetime; (2) for known or suspected carcinogens, concentration levels that represent an excess upper-bound lifetime cancer risk to an individual of between 10^{-4} and 10^{-7}; (3) other factors related to exposure (such as multiple contaminants at a site or multiple exposure pathways) or technical limitations (such as detection/quantification limits for contaminants).

ARARs define the cleanup goals when they set an acceptable level with respect to site-specific factors. As an example, Maximum Contaminant Levels (MCLs) under the SDWA are normally acceptable levels for specific contaminants. However, ARARs may not exist for some substances, or an ARAR alone would not be sufficiently protective in the given circumstances (e.g., additive effects from several chemicals that are involved). For these situations, cleanup goals may have to be based on nonpromulgated criteria and advisories (e.g., health advisories such as reference doses (R_fD)), TBCs, and state criteria, advisories, and guidance (U.S. EPA 1988).

3.4.2 RCRA REQUIREMENTS FOR HAZARDOUS WASTE AND GROUNDWATER

The RCRA requirements for treatment, storage, or disposal of hazardous wastes apply to a Superfund site if the site contains RCRA-listed or characteristic hazardous waste that was treated or disposed of, or if the CERCLA activity at the site constitutes current treatment, storage, or disposal of RCRA hazardous waste. RCRA contains several authorities under which corrective action requirements would be promulgated. Due to the similarity of corrective action under RCRA to CERCLA cleanup, these requirements are likely to be applicable or relevant and appropriate in many remedial action situations (U.S. EPA 1988).

RCRA contains groundwater monitoring and protection standards. CERCLA's goal is to restore groundwater to its beneficial uses based in large part on its vulnerability, use, and value. In general, EPA uses MCLs as protection levels for groundwater that is currently or potentially used for drinking. At particular sites where the groundwater cannot be used for drinking or where cleanup is not practicable or cost-effective, site-specific exposure-based Alternate Concentration Limits (ACLs) may be established. The *Ground-Water Protection Strategy* (U.S. EPA 1984) and draft *Office of Ground-Water Protection Classification Guidelines* (U.S. EPA 1986) serve as useful guidance.

3.4.3 Clean Water Act Requirements for Surface Water

The CWA established the basic framework for regulating discharges of pollutants into the water bodies of the United States and regulating quality standards for surface waters. Under the CWA, EPA has implemented pollution control programs such as setting wastewater standards for industry and also developed national water quality criteria recommendations for pollutants in surface waters. Unless a National Pollutant Discharge Elimination System (NPDES) permit is obtained, it is unlawful to discharge pollutants to waters of the United States.

Both on-site and off-site direct discharges from CERCLA sites to surface waters are required to meet the substantive requirements of the NPDES program. These substantive requirements include discharge limitations (both technology and water-quality based), certain monitoring requirements, and best management practices (BMPs) (U.S. EPA 1988).

3.4.4 Safe Drinking Water Act Requirements for Groundwater/Surface Water

The MCLs set under the SDWA are generally the applicable or relevant and appropriate standard for cleanup of groundwater or surface water that is, or may be, used for drinking. A standard for drinking water more stringent than an MCL may be needed in special circumstances (e.g., the presence of multiple contaminants in groundwater or the presence of extraordinary risks due to multiple pathways of exposure). For such cases, Maximum Contaminant Level Goals (MCLGs), the agency's policy on the use of appropriate risk ranges for carcinogens, levels of quantification, and other pertinent guidelines, should be considered in setting up the cleanup goals.

3.4.5 Clean Air Act Requirements for Air

Remedial activities during a CERCLA cleanup may be sources of air emissions of gas or particulate matter. Examples include handling of contaminated soil, soil vapor extraction of contaminated soil, air stripping of contaminated groundwater, thermal destruction of contaminated sludge and air, as well as bioremediation of contaminated soil and groundwater.

The CAA is to protect and enhance the ambient air quality. The potential ARARs relevant to the CAA include National Ambient Air Quality Standards (NAAQS), National Emission Standards for Hazardous Air Pollutants (NESHAPs), and New Source Performance Standards (NSPS). The six criteria pollutants under the NAAQS are carbon monoxide, lead, nitrogen dioxide, particulate matter (PM_{10} and $PM_{2.5}$), ozone, and sulfur oxides. The original list under NESHAPs included 189 hazardous air pollutants (HAPs); since 1990 the EPA has modified the list through rulemaking to 187 HAPs. The majority of the 187 HAPs are volatile organic compounds (VOCs), which are common soil and groundwater pollutants. The purpose of NSPS is to ensure that new stationary sources are designed, built, equipped, operated, and maintained to reduce emissions to a minimum (U.S. EPA 1989). In addition to federal regulations, a number of state air pollution control agencies have adopted programs to regulate toxic air pollutants.

3.4.6 RCRA REQUIREMENTS FOR AIR

RCRA regulations covering hazardous waste air emissions include (1) controls on incinerators; (2) requirements for controlling windblown fugitive PM from landfills, waste pipes, and treatment facilities; and (3) organic air emissions from TSDFs (U.S. EPA 1989).

3.4.7 TOXIC SUBSTANCE CONTROL ACT (TSCA) REQUIREMENTS

Under the Toxic Substances Control Act (TSCA), the EPA evaluates potential risks from new and existing chemicals and acts to address any unreasonable risks that chemicals may have on human health and the environment. Of these, the regulations controlling hazardous chemicals are potential ARARs for CERCLA actions. For example, the EPA has published regulations or is taking actions pertaining to polychlorinated biphenyls (PCBs), per- and polyfluoroalkyl substances (PFAS), lead, mercury, and asbestos.

3.4.8 OTHER POTENTIAL ARARs

Other potential ARARs for CERCLA actions are requirements of the Federal Insecticide, Fungicide, and Rodenticide Act, National Environmental Policy Act (NEPA), National Historic Preservation Act, Archeological and Historic Preservation Act, Endangered Species Act, Wild and Scenic Rivers Act, Fish and Wildlife Coordination Act, Coastal Zone Management Act, Wildness Act, Surface Mining Control and Reclamation Act, and standards for the cleanup of radioactively contaminated sites and buildings (U.S. EPA 1989).

3.5 RISK ASSESSMENT AND RISK MANAGEMENT

The EPA considers risk as the chance of harmful effects to human health or to ecological systems resulting from exposure to an environmental stressor. A stressor is any physical, chemical, or biological entity that can induce an adverse response. Stressors may adversely affect specific natural resources or entire ecosystems, including plants and animals, as well as the environment with which they interact. It should be noted that this section is mainly based on the information on the EPA's webpages on risk assessment (https://www.epa.gov/risk).

Environmental risk assessments are often conducted to characterize the nature and magnitude of risks from chemical contaminants and other stressors to human health and ecological receptors (e.g., plants, birds, fish, and wildlife) now or in the future. In general, the level of risk depends on the following three factors:

1. How much of a COC is present in an environmental medium (i.e., soil, water, air)?
2. How much contact/exposure does a person or ecological contactor have with the contaminated medium?
3. What is the inherent toxicity of the COC?

For site remediation, risk assessment is used to determine a safe level for humans and ecological receptors to each potentially dangerous contaminant present. It produces estimates of current and possible future risks, if a "no-action" alternative (i.e., no cleanup actions) were chosen. It can also be used to select the best cleanup alternative to manage risks (to humans and ecological receptors) to acceptable levels. "Risk-based" remediation is a commonly adopted approach for site remediation.

For a risk assessment, the following four steps should be conducted after a planning and scoping stage (Figure 3.1):

1. Hazard identification
2. Dose-response assessment
3. Exposure assessment
4. Risk characterization

Although the general approach mentioned above is applicable to both human health and ecological risk assessments, the discussion below is mainly focused on human health risk assessment.

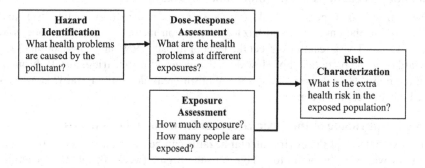

FIGURE 3.1   The process flow diagram of human health risk assessment.

3.5.1 PLANNING AND SCOPING

This planning and scoping stage is to determine the purpose and scope of a risk assessment. A risk assessment can be either qualitative (presence or absence of risks) or quantitative (level of risks). This stage also determines the technical approaches to be used. Below are the parameters that should be taken into consideration.

Receptors

Will it be an individual or general public? Are there any sensitive receptors (e.g., children, pregnant women, elderly persons, or patients)? Are there highly exposed subgroups (e.g., based on geographic area, gender, racial or ethnic group, or economic status)?

Environmental Hazards

Potential environmental hazards include chemicals, radiation, physical (e.g., dust and heat), biological/microbiological, nutritional (e.g., diet, fitness, or metabolic state), and socioeconomic (e.g., access to health care) hazards.

Sources of Environmental Hazards

Sources of the environmental hazards can be grouped into point (e.g., a contaminated site; air or water discharge from a factory), nonpoint (e.g., stormwater runoff; automobile exhausts), or natural sources (e.g., forest fires).

Exposure Pathways and Routes

Pathways through which environmental hazards can reach the receptors include air, water (i.e., surface water and groundwater), solid waste/soil, food, and nonfood consumer products/pharmaceuticals. The exposure routes include ingestion (i.e., food, water, or nondietary), inhalation, and dermal contact. Figure 3.2 illustrates a schematic of exposure pathways. The drums on the left are the source of contamination, and the environmental media include air, soil, groundwater, and biota. There are many exposure points, and the exposure routes include inhalation, ingestion, and dermal contact (not shown in the figure).

The Body's Reactions to Environmental Hazards

The reactions of a person's body (i.e., absorption, distribution, metabolism, and excretion) to the environmental hazards depend on factors such as age, race, sex, genetics, etc. These reactions affect the level of impact. How much absorption of the hazard does the body take up? How does the hazard travel/distribute itself throughout the body? How does the body's metabolism break down the hazard? How does the body excrete the hazard?

Length and Timing of the Toxic Effects of Environmental Hazards

The toxic effects of the environmental hazards can be grouped into (1) acute (right away or within a few hours to a day); (2) sub-chronic (weeks or months, generally less than 10% of the lifespan of humans); (3) chronic (a significant part of a lifetime

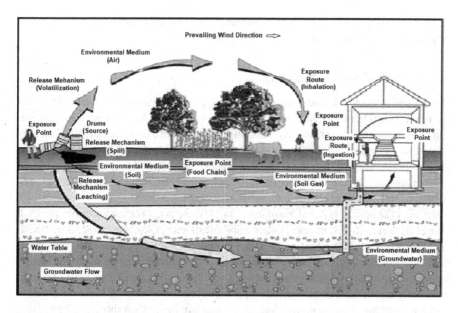

FIGURE 3.2 Schematic of exposure pathways. (Modified from ATSDR 2005.)

or a lifetime, at least seven years for humans); and (4) intermittent. In addition, some chemicals may be more toxic in a critical period of a lifetime (e.g., fetal development, childhood, and/or during aging).

Health Effects

Environmental hazards can cause death and different types of diseases of the heart, liver, and nervous system. They may also cause genetic changes, including mutagenic, carcinogenic, and teratogenic. A mutagen is a physical or chemical agent that changes the genetic material of an organism so that the frequency of mutation would increase above the natural background level. Since many mutations can cause cancer, mutagens are likely to be carcinogens, but not necessarily so. Carcinogens cause cancer. Teratogens are agents that cause abnormalities in developing organisms in the womb. When a fetal abnormality is manifested, the infant could be born with a congenital defect, anomaly, or malformation.

3.5.2 Hazard Identification

Hazard identification is the first step of risk assessment. It is a process to determine if the exposure to a stressor can cause adverse human health effects. It then determines if such an exposure can cause an increase in the incidence of specific adverse health effects, such as diseases, cancers, birth defects, and deaths. For chemical stressors, the process examines the available scientific data for a given chemical and develops and evaluates its potential to cause adverse human health effects. With regard to carcinogenicity, EPA places chemicals into five "weight of evidence" categories:

(a) human carcinogen, (b) probable carcinogen, (c) possible carcinogen, (d) not clas-
sifiable, and (e) evidence of noncarcinogenicity.

3.5.3   Dose-Response Assessment

The second step of risk assessment is a dose-response assessment, which determines
the relationship between the dose and the adverse effect. A dose-response relationship describes how the amount and condition of exposure to a stressor (the dose) are
related to the likelihood and severity of adverse health effects (the responses). The
dose-response relationship depends on the agent, the type of response (e.g., cancer,
disease, and death), the receptors (e.g., the general public, sensitive populations, and
animals), and exposure routes (e.g., inhalation and ingestion).

In general, the measured response would increase with an increase in dose.
However, there may be no (or no observable) response at low doses for some stressors. The dose at which a response begins to appear is referred to as the "threshold"
dose. The dose-response relationship between a human and a chemical should ideally
be derived from actual data involving human subjects. However, this type of data is
frequently unavailable or covers only a portion of the dose range. Often extrapolation
needs to be done to estimate the dose levels that are lower than those available from
scientific studies. In addition, animal studies are frequently conducted to augment
the available data from human objects. These doses are usually high. Extrapolation
of dose-response relationship results from animal studies to humans introduces
uncertainty.

A dose-response curve graphically represents the relationship between the dose
of a chemical and the response drawn. To include extrapolation, the dose-response
relationship can be classified into two types: nonlinear and linear. For a nonlinear
dose-response relationship, the toxicity has a threshold; it means that no adverse
effect is expected to occur when the dose is below a finite value (see Figure 3.3a). A
No-Observed-Adverse-Effect (NOAEL) is the highest experimentally determined
dose that would not cause a statistically- or biologically-significant adverse effect. For

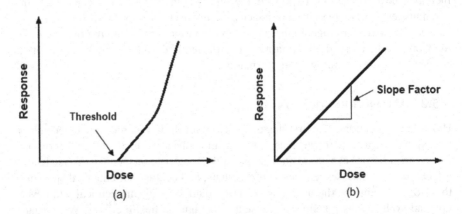

FIGURE 3.3   Types of dose-response relationships: (a) nonlinear, and (b) linear.

cases that a NOAEL has not been experimentally demonstrated, the term "lowest-observed-adverse-effect level" (LOAEL) is used to describe the lowest dose tested.

The reference dose (R_fD) is defined as an estimate of a daily oral exposure to the human population (including the sensitive receptors) that is likely to be without an appreciable risk of harmful effects during a lifetime. It is often derived from NOAEL or LOAEL with order-of-magnitude uncertainty factors (UFs) or safety factors. It is generally expressed in units of milligrams per kilogram of body weight per day (mg/kg/day). A similar term, known as reference concentration (R_fC), is used to assess inhalation risks, where concentration refers to levels in the air (generally expressed in units of milligrams of the chemical agent per cubic meter of air, mg/m^3). For noncancer health effects, risk from exposure to a specific quantity is expressed as a *Hazard Index* (HI), which is the average daily dose (ADD) divided by R_fD:

$$\text{Hazard Index} \left(\text{HI}\right) = \text{Average Daily Dose} \left(\text{ADD}\right) \div R_fD \qquad (3.1)$$

No adverse effects are expected if values of HI are less than 1. The total HI vale for an exposure scenario is the sum of the HI for each contaminant.

If the toxicity does not have a threshold value, a linear dose-response relationship is typically used. This relationship is generally assumed for carcinogens. The slope of the straight line is called slope factor, or cancer slope factor (see Figure 3.3b). In other words, the factor is the ratio of risk and dose, which is derived based on an assumption that every dose poses a risk and that there is no safe dose (i.e., no threshold values). Table 3.2 tabulates the oral and/or inhalation cancer slope factors of some carcinogens. When linear dose-response is used to assess cancer risk, the excess lifetime cancer risk resulting from exposure to a chemical agent can be calculated as:

$$\text{Cancer Risk} = \text{Exposure} \times \text{Cancer Slope Factor} \qquad (3.2)$$

Total cancer risk is then calculated by summing the individual excess cancer risks of each pollutant in each exposure pathway (i.e., inhalation, ingestion, and dermal absorption). A more detailed discussion can be found in EPA's *Guidelines for Carcinogen Risk Assessment* (U.S. EPA 2005).

Example 3.1: Calculating Cancer Risk Using the Slope Factor

The exposure to a contaminated site results in a person receiving an average daily oral dose of 0.0001 mg benzene per day per kilogram of body weight. Determine the person's excess life cancer risk from this exposure.

SOLUTION:

Using Eq. 3.2, Cancer Risk = Exposure × Cancer Slope Factor

$$= \left(0.0001 \text{ mg/d/kg}\right) \times \left(0.055/\text{mg/d/kg}\right) = 5.5 \times 10^{-6} \left(\text{or 5.5 in 1 million}\right)$$

TABLE 3.2
Cancer Slope Factors of Some Carcinogens

Chemical Name	CAS Number	Oral Slope Factor (1/mg/kg/d)	Inhalation Slope Factor (1/mg/kg/d)
Alachor	15972608	0.08	
Arsenic	7440382	1.5	15.1
Benzene	71432	0.055	0.027
Bromoform	75252	0.0079	0.0039
Carbon tetrachloride	56235	0.13	0.053
Chlordane	57749	0.35	0.35
Chloroethane	75003	0.0029	
Chloroform	67663		0.081
1,4-Dichlorobenzene	106467	0.024	0.022
1,2-Dichloroethane	107062	0.091	0.091
1,3-Dichloropropene	542756	0.1	0.01
Heptachlor	76448	4.5	4.5
Hexachlorobenzene	118741	1.6	1.6
Hexachloroethane	67721	0.014	0.014
Methylene chloride	75092	0.0075	0.00165
1,1,1,2-Tetrachoroethane	630206	0.026	0.026
1,1,2,2-Tetrachloroethane	79345	0.2	0.2
Tetrachloroethylene	127184	0.54	0.02
1,1,2-Trichloroethane	79005	0.057	0.056
Trichloroethylene	79016	0.4	0.4
2,4,6-Trichlorophenol	88062	0.011	0.01

DISCUSSION:

1. Since it is the oral dose, the oral slope factor (Table 3.2) was used in the calculation.
2. For the EPA, an excess lifetime risk of 1 in 10,000 (or 100 in 1 million) to 1 in 1,000,000 is often considered acceptable. Individual states may have different criteria.

3.5.4 EXPOSURE ASSESSMENT

The exposure assessment is the process of determining the magnitude, frequency, and duration of human exposure to an agent in the environment. The assessment can be done to estimate the past, current, or future exposure for an agent. In a human-health risk assessment, the exposure assessment step should be conducted in parallel with the dose-response relationship step.

An exposure assessment should consider the size, nature, and types of human populations exposed to the agent. In addition, exposure assessment should consider the exposure pathway/media (i.e., water, air, and soil) as well as the exposure routes (i.e., ingestion, inhalation, and eye/dermal contact). In addition to having direct measurements of the agent concentrations in different media that a receptor is receiving, a few models may need to be used, including the release model from a contaminant source and fate and transport model of the agent through each medium. More detailed information can be found in EPA's *Guidelines for Human Exposure Assessment* (U.S. EPA 2019).

Eq. 3.2 is a generic equation to estimate the exposure dose resulting from contact with a contaminated medium.

$$\text{Average Daily Dose}\left(\text{ADD}\right) = \frac{(C)(IR)(AF)(EF)}{(BW)} \tag{3.3}$$

where ADD=average daily dose; C=concentration; IR=intake rate; AF=absorption factor; EF=exposure factor; and BW=body weight (kg).

The *absorption factor* (AF), or *bioavailability factor*, represents the percentage of the total amount of a substance ingested, inhaled, or contacted that actually enters the bloodstream and is available to possibly harm a person. It is typically assumed to be 1 (100%), meaning all of a substance to which a person is exposed is absorbed, as a conservative approach, unless more scientific data are available.

For a continuous exposure to the contaminant, the exposure factor (EF) would be equal to 1. However, some exposure may occur on an irregular or intermittent basis. For these cases, the EF is calculated by multiplying frequency of exposure (FE), in days/year, by the exposure duration (ED), in years, and then dividing by the averaging time (AT), which is the time period in which the dose is to be averaged, in days, as shown in Example 3.2. If the chemical is not a carcinogen, the AT is the duration of exposure. If the chemical is a carcinogen, the AT should be a lifetime, typically 70 years.

$$\text{Exposure Factor}\left(\text{EF}\right) = \frac{(FE)(ED)}{(AT)} \tag{3.4}$$

Example 3.2: Exposure Factor

Determine the exposure factor for a child who came into contact with contaminated soil three days per week, 40 weeks per year, during a 4-year period.

SOLUTION:

Using Eq. 3.4, Exposure Factor=(FE)(ED) ÷ AT
   = {[(3 days/week) × (40 weeks/year)] × 4 years} ÷ [(4 years) × (365 days/year)]
   = 0.33

By inserting Eq. 3.4 into Eq. 3.3, the generic dose equation becomes Eq. 3.5 as shown below:

$$\text{Average Daily Dose}\left(\text{ADD}\right) = \frac{(C)(IR)(AF)(FE)(ED)}{(AT)(BW)} \tag{3.5}$$

The values used in Eq. 3.5 should depend on site-specific exposure conditions. EPA's *Exposure Factors Handbook* (U.S. EPA 1997) is a good source of exposure information that may be relevant to the site being evaluated. If site-specific information is not available, conservative assumptions should be used.

Some standard default values are also commonly used. For example, default values for body weight (BW) are 10, 16, and 70 kg for infants (6–11 months), children 1–6 years old, and adults, respectively. Those for exposure duration (ED) are 9 and 70 years for the national median time at one residence and lifetime, respectively. Default average drinking water intake rates (IR) are 1 and 2 L/day for children and adults, respectively. Default average soil ingestion rates are 200 and 100 mg/day for children and adults, respectively. Default air intake rates are 10, 11.3, and 15.2 m³/day for 6–8-year-old children, adult females, and adult males, respectively.

Example 3.3: Exposure Doses from Water Ingestion

A water supply is contaminated with 10 mg/L of methylene chloride. Use the default values to calculate the average daily ingestion exposure doses for (a) adults and (b) children 1–6 years old.

SOLUTION:

(a) For an adult, use Eq.3.3 to calculate the average daily dose:

$$ADD = \frac{(C)(IR)(AF)(EF)}{(BW)} = \frac{\left(10^{mg}/_L\right)\left(2\,^L/_d\right)(1)(1)}{70\,kg} = 0.286\,\frac{mg}{d}\,/kg$$

(b) For children, also use Eq.3.3 to calculate the average daily dose:

$$ADD = \frac{(C)(IR)(AF)(EF)}{(BW)} = \frac{\left(10^{mg}/_L\right)\left(1\,^L/_d\right)(1)(1)}{16\,kg} = 0.625\,\frac{mg}{d}\,/kg$$

DISCUSSION:

1. The units of concentration (C) and intake rate (IR) should match to yield units of mg/d for the multiplication product of C and IR.
2. Without additional scientific data and for the worst-case scenario, the absorption factor (AF) is assumed to be 1 for both cases.
3. Without additional scientific data and for the worst-case scenario, the exposure factor (EF) is assumed to be 1 for both cases.
4. Although the water ingestion rate of children is only half of that of adults, the average daily dose of children is higher because of the smaller body weight (16 kg) used in the calculations.

Example 3.4: Exposure Dose from Soil Ingestion

The soil at a site is contaminated with a noncarcinogenic chemical concentration of 100 mg/kg. Assuming a child is on-site 3 days per week, 40 weeks per year, for a period of 4 years, calculate the exposure dose from soil ingestion.

SOLUTION:

From Example 3.2, the exposure factor = 0.33.
Use Eq.3.3 to calculate the average daily dose:

$$ADD = \frac{(C)(IR)(AF)(EF)}{(BW)} = \frac{\left(100\,\frac{mg}{kg}\right)\left(200\,\frac{mg}{d} \times \frac{10^{-6}kg}{mg}\right)(1)(0.33)}{16\ kg}$$

$$= 0.625\,\frac{mg}{d}\,/kg$$

DISCUSSION:

1. The units of concentration (C) and intake rate (IR) should match to yield units of mg/d for the multiplication product of C and IR.
2. Without additional scientific data and for the worst-case scenario, the absorption factor (AF) is assumed to be 1.

Example 3.5: Exposure Dose from Soil Dermal Contact

The soil at a site is contaminated with a noncarcinogenic chemical concentration of 100 mg/kg. Assuming the absorption factor (or bioavailability factor) is equal to 0.1, calculate the average daily exposure dose for a child that has been exposed to this contaminated soil 200 days per year from birth through 6 years of age. The default amounts of soil adhered are 210, 525, 299, and 326 mg/day for age groups of 0–1, 1–11, 12–17, and 17–70 years, respectively.

SOLUTION:

(a) The exposure factor for age 0–1 = (FE)(ED) ÷ AT
= (200 days/year × 1 year) ÷ (6 years × 365 days/year) = 0.091
(b) The exposure factor for age 1–6 = (FE)(ED) ÷ AT
= (200 days/year × 5 years) ÷ (6 years × 365 days/year) = 0.46

(c) $ADD = \dfrac{(C)(IR)(AF)(EF)}{(BW)}$

$$= \frac{\left(100\,\frac{mg}{kg}\right)\left(210\,\frac{mg}{d} \times \frac{10^{-6}kg}{mg}\right)(0.1)(0.091)}{10\,kg}$$

$$+ \frac{\left(100\,\frac{mg}{kg}\right)\left(525\,\frac{mg}{d} \times \frac{10^{-6}kg}{mg}\right)(0.1)(0.46)}{16\,kg}$$

$$= 0.00017\,\frac{mg}{d}\,/kg$$

Discussion: Note that from age 0 to 1, the averaging time (AT) used was 6 years. This is because the total time of exposure is 6 years. Then, when calculating ADD, we must consider first the 0–1 year period and then 1–6 years because the soil adsorption rate is different for those two age groups.

Example 3.6: Exposure Dose from Inhalation

The existing National Ambient Air Quality Standard (NAAQS) of PM_{10} is 150 µg/m³ (on an average of 24 hours). Estimate the daily average dose of PM_{10} for men and women, if the PM_{10} at a location is always at the NAAQS.

SOLUTION:

(a) For men, $ADD = \dfrac{(C)(IR)(AF)(EF)}{(BW)} = \dfrac{\left(150\frac{\mu g}{m^3} \times \frac{10^{-3}mg}{\mu g}\right)\left(15.2\frac{m^3}{d}\right)(1.0)(1.0)}{70 \text{ kg}}$

$$= 0.033 \frac{mg}{d}/kg$$

(b) For women, $ADD = \dfrac{(C)(IR)(AF)(EF)}{(BW)} = \dfrac{\left(150\frac{mg}{m^3} \times \frac{10^{-3}mg}{\mu g}\right)\left(11.3\frac{m^3}{d}\right)(1.0)(1.0)}{70 \text{ kg}}$

$$= 0.024 \frac{mg}{d}/kg$$

DISCUSSION:

1. For a conservative approach, these assumptions are used: (1) PM_{10} concentration is the same as that of NAAQS; (2) absorption factor (AF)=1; and (3) exposure factor (EF)=1.
2. The ADD values are different for men and women because of the different air intake rates.

Example 3.7: Hazard Index and Reference Dose

A city has groundwater as the potable water supply. The aquifer is contaminated with toluene. The oral reference dose of toluene is 0.2 mg/kg/day. By taking a safety factor of 20, the allowable average daily dose is set at 0.01 mg/kg/day. At this daily dose, (a) what would be the hazard index? and (b) what would be the maximum allowable toluene concentration in the drinking water to protect adults and children?

SOLUTION:

(a) Use Eq. 3.1, Hazard Index (HI) = ADD ÷ R_fD = 0.01 ÷ 0.2 = 0.05
(b) For adults, use Eq.3.3 to calculate the maximum allowable concentration:

$$0.01\frac{mg}{d}/kg = \frac{(C)(IR)(AF)(EF)}{(BW)} = \frac{(C)(2\,L/d)(1)(1)}{70 kg} \qquad C = 0.35 \text{ mg/L}$$

(c) For children, also use Eq.3.3 to calculate the maximum allowable concentration:

$$0.01\frac{mg}{d}/kg = \frac{(C)(IR)(AF)(EF)}{(BW)} = \frac{(C)(1\,L/d)(1)(1)}{16 kg} \qquad C = 0.16 \text{ mg/L}$$

Discussion:

1. By setting the allowable average daily dose smaller than the R_fD, the HI becomes less than 1, which is more acceptable.
2. As a worst-case scenario, both the absorption and exposure factors are set to be 1.
3. Although the average daily doses for adults and children are the same at 0.01 mg/kg/d, the allowable concentrations are different (0.35 vs. 0.16 mg/L). The lower concentration should be used as the compliance limit.

3.5.5 RISK CHARACTERIZATION AND BEYOND

Risk characterization is to summarize and integrate information from the previous steps to synthesize an overall conclusion about the risk. EPA's *Risk Characterization Handbook* (U.S. EPA 2000) provides details on risk characterization. Risk characterization is the final assessment process and the first input to the risk management process.

To summarize, risk assessment collects information, makes assumptions, and generates information on potential health or ecological risks. Environmental risk management is to determine how to manage the risks, based on the information obtained from risk assessment, in a way best suited to protect human health and the environment by also taking other factors (e.g., economic and social factors) into consideration. For site remediation, risk communication, which is a process of informing people about potential hazards to them, their property, and/or their community, should be well-practiced.

3.6 SUMMARY

This chapter described the major environmental laws and regulations pertaining to site assessment and remediation, in addition to discussing risk assessment calculations. Those concepts all have the goal to protect human health and the environment. The major concepts covered were:

- Comprehensive Environmental Response, Compensation, and Liability Act (CERCLA), or Superfund
- Resource Conservation and Recovery Act (RCRA)
- Applicable or Relevant and Appropriate Requirements (ARARs)
- Parameters taken into consideration for planning and scoping a risk assessment: receptors, environmental hazards, sources of environmental hazards, exposure pathways (air, water, solid waste, food, and consumer products), and routes (ingestion, inhalation, and dermal contact).
- Hazard identification, including dose-response assessment, hazard index (HI), exposure assessment, average daily dose (ADD), and exposure factor (ED).

3.7 PROBLEMS AND ACTIVITIES

3.1. The exposure to a contaminated site results in a person receiving average daily oral and inhalation doses, each, of 0.00001 mg benzene per day per kilogram of body weight. Determine the excess life cancer risk from this oral and inhalation exposure.

3.2. Determine the exposure factor for a child who came into contact with contaminated soil 3 days a week, 30 weeks per year, during a 6-year period.

3.3. A water supply is contaminated with 0.5 mg/L of benzene. Estimate the average daily ingestion exposure doses for (a) adults and (b) children 1–6 years old.

3.4. The soil at a site is contaminated with a noncarcinogenic concentration of 100 mg/kg. Assuming an adult is on site 5 days a week, 50 weeks per year, for a period of 3 years, calculate the exposure dose from soil ingestion.

3.5. The soil at a site is contaminated with a noncarcinogenic concentration of 100 mg/kg. Assuming the absorption factor (or bioavailability factor) is equal to 0.1, calculate the average daily exposure dose for an adult who has been exposed to this contaminated soil 200 days per year for 5 years.

3.6. The existing National Ambient Air Quality Standard (NAAQS) of $PM_{2.5}$ is 35 μg/m^3 (24 hours average). Estimate the daily average dose of $PM_{2.5}$ for a 7-year-old child if the $PM_{2.5}$ at a location is always below the corresponding NAAQS.

3.7. A city has groundwater as the potable water supply. The aquifer is contaminated with phenol. The oral reference dose of phenol is 0.6 mg/kg/day. By taking a safety factor of 50 as the allowable average daily dose, (a) what would be the hazard index? (b) what would be the maximum allowable phenol concentration in the drinking water to protect children and adults?

REFERENCES

ATSDR. (2005). *Public Health Assessment Guidance Manual (2005 update)*. Agency for Toxic Substances and Disease Registry (ATSDR), U.S. Department of Health and Human Services, Atlanta, GA.

U.S. EPA. (1984). *Ground-Water Protection Strategy*. U.S. Environmental Protection Agency Office of Ground-Water Protection, Washington, DC.

U.S. EPA. (1986). *Guidelines for Ground-Water Classification Under the EPA Ground-Water Protection Strategy -- Final Draft*. U.S. Environmental Protection Agency Office of Ground-Water Protection, Washington, DC.

U.S. EPA. (1988). *CERCLA Compliance with Other Laws Manual: Interim Final*. EPA/540/G-89/006. U.S. Environmental Protection Agency Office of Solid Waste and Emergency Response, Washington, DC.

U.S. EPA. (1989). *CERCLA Compliance with Other Laws Manual: Part II. Clean Air Act and Other Environmental Statutes and State Requirements*. EPA/540/G-89/009. U.S. Environmental Protection Agency Office of Emergency and Remedial Response, Washington, DC.

U.S. EPA. (1994). *Feasibility Study Analysis for CERCLA Municipal Landfill Sites*. EPA/540/R-94/081. U.S. Environmental Protection Agency Office of Solid Waste and Emergency Response, Washington, DC.

U.S. EPA. (1997). *Exposure Factors Handbook*. U.S. Environmental Protection Agency Office of Research and Development, Washington, DC.

U.S. EPA. (2000). *Risk Characterization Handbook*. EPA/100/B-00/002. U.S. Environmental Protection Agency Science Policy Council, Washington, D.C.

U.S. EPA. (2005). *Guidelines for Carcinogen Risk Assessment*. EPA/630/P-03/001F. U.S. Environmental Protection Agency Risk Assessment Forum, Washington, DC.

U.S. EPA. (2011). *This is Superfund – A Community Guide to EPA's Superfund Program*. EPA/540/R-11/021. U.S. Environmental Protection Agency Office of Solid Waste and Emergency Response, Washington, DC.

U.S. EPA. (2014). *RCRA Orientation Manual 2014. Resource Conservation and Recovery Act*, EPA/530/F-11/003. U.S. Environmental Protection Agency Office of Resource Conservation and Recovery, Washington, DC.

U.S. EPA. (2018). *Hazardous Waste Treatment, Storage, and Disposal Facilities (TSDF) Regulations*. EPA/530/F-11/006 (Version 8). U.S. Environmental Protection Agency Office of Resource Conservation and Recovery, Washington, DC.

U.S. EPA. (2019). *Guidelines for Human Exposure Assessment*. EPA/100/B-19/001. U.S. Environmental Protection Agency Risk Assessment Forum, Washington, DC.

4 Site Assessment and Remedial Investigations

4.1 INTRODUCTION TO SITE ASSESSMENT AND REMEDIAL INVESTIGATION

Our understanding and knowledge in conducting thorough site assessments (SA) and remedial investigations (RI) have matured over the past decade, in which sound scientific principles have been better used to assure appropriate "due diligence" in performing such studies. Due diligence is the reasonable care/action that a reasonable person is expected to take to avoid harm to others or their property. There are numerous state and federal guidelines and technical documents to fill the gap in our understanding of what to do. Adoption of formal procedures that help guide one through the process is best left to the environmental professionals to assure that due diligence follows in the footsteps of one's state and/or federal regulatory framework. Generally, activities in SA/RI include contaminant characterization; source identification; field techniques to sample soil vapor, soil matrices, and groundwater; determination of site conditions to develop a site conceptual model (SCM); and an understanding of pathways to sensitive environmental receptors. An SCM is an iterative and living representation of a site that summarizes the available and essential data/information to a guiding project team through the entire site cleanup process. A good SCM synthesizes chemical data with geological, hydrogeological, and other site information to enhance the project team's ability to develop solutions to ensure the protection of human health and the environment through the effective management of resources while limiting the environmental footprint of site cleanup activities (U.S. EPA 2011). Figures 4.1 and 3.1 illustrate parameters that compose SCMs, including contamination source, release mechanism, environmental media and transport, exposure points, and exposure routes.

4.1.1 ENVIRONMENTAL SITE ASSESSMENTS, PHASE I, II, AND III

The first part of an environmental site assessment (ESA), otherwise referred to as a Phase I site assessment, is generally considered as the first step in the process of environmental due diligence. Standards for performing an ESA have been promulgated by the EPA and are based partly on ASTM Standard E1527-13 – *Standard Practice for Environmental Site Assessments: Phase I Environmental Site Assessment Process*. Common tasks of a typical Phase I site assessment include:

- A review of historical records of the property, including aerial photographs, fire insurance mapping, and topographical mapping

- A review of regulatory records of the property such as hazardous waste manifests, as well as spills and releases to the environmental media (e.g., Title III of SARA) as well as regulatory permits such as wastewater discharge permits, air discharge permits, and Spill Prevention Control and Countermeasure Plan (SPCCP)
- Interviews with owners, occupants, and other individuals with regard to property history, property use, and environmental issues
- A site reconnaissance to identify present and past uses and environmental conditions

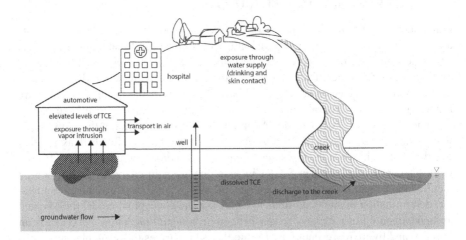

FIGURE 4.1   A site conceptual model showing the chemical trichloroethylene (TCE) originating at the automotive facility. Humans and the environment are exposed to TCE in several ways: inhalation via TCE vapor intrusion into the building; ingestion and skin contact via groundwater that is pumped through the drinking water well and distributed to the population; and aquatic life exposure via TCE-laden groundwater that interacts with the creek.

In short, a Phase I site assessment consists of reviewing records and interviewing individuals. If the Phase I site assessment determines a likelihood of site contamination, then a Phase II site assessment is usually conducted. A Phase II assessment is an "intrusive" investigation in which samples of soil and/or groundwater are collected and analyzed for quantitative values of the contaminants. Guidelines for a Phase II site assessment can be found in ASTM E1903-19 – *Standard Practice for Environmental Site Assessments: Phase II Environmental Site Assessment Process.*

A Phase III site assessment is an investigation involving site remediation. The purpose of a Phase III site assessment is to delineate the physical extent of contamination based on recommendations made in the Phase II assessment. Potential activities may include intensive testing, sampling, monitoring, and conducting "fate and transport" studies as well as feasibility studies for remediation plans. The findings are used to determine the steps needed to perform site cleanup and the follow-up monitoring for residual contaminants.

This phase is usually called remedial investigation (RI). RI activities consist of additional site characterization and collection of data needed to make engineering decisions on the control of plume migration and selection of remedial alternatives. The questions to be answered by the RI activities typically include the following:

- What media (surface soil, vadose zone, underlying aquifer, air) have been impacted?
- Where is the plume located in each impacted medium?
- What are the vertical and areal extents of the plume?
- What are the concentration levels of compounds of concern (COCs)?
- How long has the plume been there?
- Where is the plume going?
- Has the plume gone beyond the property boundary?
- How fast will the plume go?
- What are the on-site sources of the COCs?
- Are there potential off-site sources to this plume (now and/or in the past)?

Subsurface contamination from spills and leaky underground storage tanks (USTs) creates environmental conditions that usually require corrective remedies. The COCs may be present in a combination of the following locations and phases, described below and shown in Figure 2.5, which is repeated here as Figure 4.2.

Vadose (unsaturated) zone

- Vapors in the soil voids
- Free product in the soil voids
- Dissolved in the soil moisture
- Adsorbed onto the soil grains
- Floating on top of the capillary fringe (for light nonaqueous phase liquid [LNAPLs])

Underlying aquifer

- Dissolved in groundwater
- Adsorbed onto the aquifer material
- Coexisting with groundwater in the pores as free product or sitting on top of the bedrock (for dense nonaqueous phase liquid [DNAPLs])

Common RI activities may include:

- Removal of the source(s) of contamination, such as leaky USTs
- Installation of soil borings
- Installation of groundwater monitoring wells
- Collection and analysis of soil samples
- Collection and analysis of groundwater samples

- Collection of groundwater elevation data
- Performance of aquifer testing
- Removal of impacted soil that may serve as a contamination source to the aquifer

Through these activities, the following data may be collected:

- Types of COCs present in the vadose zone and underlying aquifer
- Concentrations of COCs in the collected soil and groundwater samples
- Vertical and areal extents of the plumes in the vadose zone and underlying aquifer
- Vertical and areal extents of the free product (LNAPLs and DNAPLs)
- Soil characteristics, including types, density, porosity, and moisture content
- Groundwater elevations
- Drawdown data from aquifer tests

Using these collected data, engineering calculations are then performed to assist in site remediation. Common engineering calculations include:

- Mass and volume of impacted soil in the vadose zone
- Mass of COCs in the vadose zone

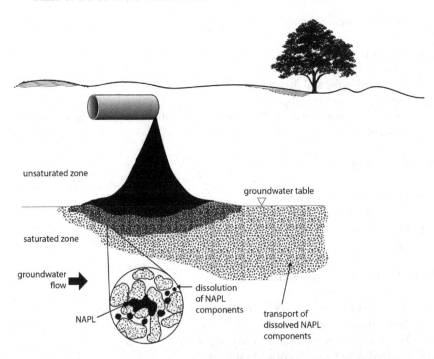

FIGURE 4.2 Spread of COCs leaking from an underground storage tank (UST), moving from the unsaturated (vadose) zone to the groundwater aquifer.

- Mass and volume of the free product (LNAPLs and DNAPLs)
- Size of the dissolved plume in the aquifer
- Mass of COCs present in the aquifer (dissolved and adsorbed)
- Hydraulic gradient and groundwater flow direction
- Hydraulic conductivity of the aquifer

This chapter, along with Chapter 2, describes most of the engineering calculations needed for the items above. The chapter will discuss calculations related to site assessment activities, including cuttings from soil borings and purge water from groundwater sampling. A good understanding of the partitioning phenomenon of COCs is critical for the evaluation of the fate and transport of COCs in the subsurface (see Chapter 2) and the selection of remedial alternatives.

4.1.2 AMOUNT OF IMPACTED SOIL IN THE VADOSE ZONE

Chemicals that leaked from USTs might move beyond the vicinity of the tank. If subsurface contamination is suspected, soil borings are drilled to assess the extent of contamination in the vadose zone. Soil boring samples are then taken at a fixed interval, e.g., every 5 or 10 ft, and analyzed for soil properties. Selected samples are submitted to certified laboratories and analyzed for COCs. Figure 4.3 shows an operating drill rig, and Figure 4.4 shows an example of soil cuttings collected from

FIGURE 4.3   A drill rig drilling a soil boring. (Photo credit: Michael Shiang.)

FIGURE 4.4 Cuttings from soil borings. (Photo credit: Michael Shiang.)

the borings. Notice the different colors and textures of the collected soils, showing a variety of soils encountered in the subsurface.

When selecting remedial alternatives, an engineer needs to know the vertical and areal extents of the plume, types of subsurface soil, types of COCs, mass and volume of the impacted soil, and mass of COCs in different phases. A useful visualization made by geologists and engineers is called a soil stratigraphy drawing, shown in Figure 4.5. In this figure, we see the locations of four soil borings, their depths, and the intervals

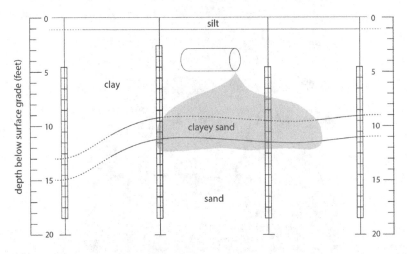

FIGURE 4.5 Soil stratigraphy drawing showing a leaking underground storage tank (UST), soil borings, interpolated soil types, and plume extent.

at which soil samples were collected. These figures are then used to interpolate the soil types and the extent of the contaminant plume in the subsurface. If the location of the plume is shallow (not deep below the ground surface level) and the amount of the impacted soil is not extensive, excavation coupled with on-site aboveground treatment (discussed in Chapter 5) or off-site treatment and disposal may be a viable option. On the other hand, *in situ* remediation alternatives, such as soil venting, would be more favorable if the volume of the impacted soil is large and deep. Therefore, a good estimate of the amount of impacted soil in the vadose zone is important for considering the remedial alternatives that are described in Chapters 5, 6, and 7.

The following procedure can be used to estimate the volume and mass of the impacted soil in the vadose zone:

Step 1: Determine the area of the plume at each sampling depth, A_i.

Step 2: Determine the thickness interval for each area calculated above, h_i.

Step 3: Determine the volume of the impacted soil, V_s, using the following formula:

$$V_s = \Sigma A_i h_i \qquad (4.1)$$

Step 4: Determine the mass of the impacted soil, M_s, by multiplying V_s with the total bulk density of soil, ρ_t, as:

$$M_s = \rho_t \times V_s \qquad (4.2)$$

To determine the mass and volume of the impacted water contained in a groundwater plume, the following procedure should be followed:

Step 1: Use Eq. 4.1 to determine the volume of the plume.

Step 2: Multiply the volume from Step 1 by the aquifer porosity to obtain the volume of the impacted groundwater.

Step 3: Multiply the volume from Step 2 by groundwater density to obtain the mass of the impacted water, if needed.

Example 4.1: Determine the Amount of Impacted Soil in the Vadose Zone

Five soil borings were installed at a site after USTs were removed. Soil samples were taken every 5 ft below ground surface (bgs). Based on the sampling results, the area of the plume at each soil sampling interval was determined as follows:

Depth (feet bgs)	Area of the plume (ft²)
15	0
20	350
25	420
30	560
35	810
40	0

The total bulk density of the soil is 112 lb/ft³. Determine the volume and mass of the impacted soil left in the vadose zone.

Strategy: The soil samples were taken and analyzed every 5 ft; therefore, each plume area represents the same depth interval. The sample taken at the 20-ft depth represents the 5-ft interval from 17.5–22.5 ft (the mid-depth of the first two consecutive intervals to the mid-depth of the next two consecutive intervals). Similarly, the sample at the 25-ft depth represents the 5-ft interval from 22.5 ft to 27.5 ft, and so on.

SOLUTION:

Thickness intervals for all areas of the plume are the same at 5 ft.

Volume of the impacted soil (using Eq. 4.1)

$$= (5 \text{ ft})(350 \text{ ft}^2) + (5 \text{ ft})(420 \text{ ft}^2) + (5 \text{ ft})(560 \text{ ft}^2) + (5 \text{ ft})(810 \text{ ft}^2)$$

$$= (1,750 + 2,100 + 2,800 + 4,050) \text{ ft}^3 = 10,700 \text{ ft}^3 = 396 \text{ yd}^3$$

or $$= (22.5 - 17.5)(350) + (27.5 - 22.5)(420) + (32.5 - 27.5)(560)$$
$$+ (37.5 - 32.5)(810) = 10,700 \text{ ft}^3$$

The mass of the impacted soil is calculated using Eq. 4.2:

$$M = (10,700 \text{ ft}^3)(112 \text{ lb/ft}^3) = 1,198,400 \text{ lb} = 599 \text{ tons}$$

Example 4.2: Determine the Amount of Impacted Soil in the Vadose Zone (in SI units)

After leaky USTs were removed from a site, five soil borings were installed. Soil samples were taken every 2 m below ground surface. However, not all the samples were analyzed due to budget constraints. Based on the analyzed samples, the areas of the plume at a few depths were determined as follows:

Depth (meter bgs)	Area of the plume at that depth (m²)
6	0
8	35
10	42
14	81
16	0

The total bulk density of soil is 1,800 kg/m³. Determine the volume and mass of the impacted soil left in the vadose zone.

Strategy: The depth intervals given are not the same; therefore, each plume area represents a different depth interval. For example, the sample collected at the 10-m depth represents a 3-m interval, from 9 to 12 m.

SOLUTION:

Volume of the impacted soil (using Eq. 4.1) = $(2)(35) + (3)(42) + (3)(81)$ m^3 = 439 m^3

$$\text{or} = (9-7)(35) + (12-9)(42) + (15-12)(81) = 439 \text{ m}^3$$

Mass of the impacted soil (using Eq. 4.2) = (439 m^3)(1,800 kg/m^3) = 790,200 kg = 790 tonnes = 870 tons.

Discussion: For the SI system of units, 1 tonne or metric ton = 1,000 kg (equivalent to 2,200 lb), and for the U.S. customary system (USCS) of units, 1 (short) ton = 2,000 lb.

4.1.3 HEIGHT OF THE CAPILLARY FRINGE

The capillary fringe (or capillary zone) is a zone immediately above the water table of an unconfined aquifer. It extends upward from the top of the water table due to the capillary rise of water. The capillary fringe often creates complications in site remediation projects. In general, the size of the plume in the aquifer would be much larger than that in the vadose zone because the dissolved plume spreads with groundwater movement. If the water table fluctuates, the capillary fringe moves upward or downward with the water table. Consequently, the capillary fringe above the dissolved groundwater plume can become impacted. In addition, if free-floating products exist, the fluctuation of the water table causes the free product to move vertically and laterally. The site remediation for this scenario is more complicated and difficult.

The height of the capillary fringe at a site strongly depends on its subsurface geology. For pure water at 20°C in a clean glass tube, the height of capillary rise is approximately:

$$h_c = \frac{0.153}{r} \tag{4.3}$$

where h_c is the height of capillary rise in centimeters, and r is the radius of the capillary tube in centimeters. Applied to soil media, h_c can be thought of as the thickness of the capillary fringe, and r can represent the pore size of the formation. This formula can be used to estimate the height of the capillary fringe. As shown in Eq. 4.3, the thickness of the capillary fringe varies inversely with the pore size of the formation. Table 4.1 summarizes the information from two references with regard to the capillary fringe. For small soil grains, the pore radius is often smaller, and the capillary rise increases. The thickness, or the height, of the capillary fringe of a clayey aquifer can be as large as 10 ft.

Example 4.3: Thickness of the Capillary Fringe

A core sample was taken from an impacted unconfined aquifer and analyzed for pore size distributions. The effective pore radius was determined to be 5 μm. Estimate the thickness of the capillary fringe of this aquifer.

SOLUTION:

Pore radius $= 5 \times 10^{-6}$ m $= 5 \times 10^{-4}$ cm
Using Eq. 4.3, capillary rise $= (0.153)/(5 \times 10^{-4}) = 306$ cm $= 3.06$ m $= 10.0$ ft

Discussion:

1. Eq. 4.3 is an empirical equation. The units for capillary rise and pore radius in this equation need to be in centimeters. By looking at this equation, having both units in centimeters does not seem to match. However, the constant (0.153) has taken care of the unit conversions. If other units are used, the value of the constant would be different.
2. The calculated value (306 cm) for the capillary rise is essentially the same as the value (300 cm) listed in Table 4.1 for clay with a pore radius of 0.005 mm.

TABLE 4.1
Typical Height of Capillary Fringe

Material	Grain Size (mm)[a]	Pore Radius (cm)[b]	Capillary Rise (cm)	
Coarse gravel		0.4		0.38[b]
Fine gravel	5–2		2.5[a]	
Very coarse sand	2–1		6.5[a]	
Coarse sand	1–0.5	0.05	13.5[a]	3.0[b]
Medium sand	0.5–0.2		24.6[a]	
Fine sand	0.2–0.1	0.02	42.8[a]	7.7[b]
Silt	0.1–0.05	0.001	105.5[a]	150[b]
Silt	0.05–0.02		200[a]	
Clay		0.0005		300[b]

Sources: [a]Todd 1980; [b]Fetter, Jr. 1980.

4.1.4 MASS AND VOLUME OF THE FREE-FLOATING PRODUCT

The LNAPL product leaked from a UST may accumulate on the top of the capillary fringe of a water-table (unconfined) aquifer instead of on top of the water table. LNAPL can also accumulate on the top of the upper confining layer of a confined aquifer to form a free-product layer. For site remediation, it is often necessary to estimate the volume or mass of this free-floating product. The thickness of the free product found in the groundwater monitoring wells had been directly used to calculate the volume of free product outside the wells (Figure 4.6 illustrates how LNAPL appears on top of the water inside a monitoring well). However, these calculated values are seldom representative of the actual free product volume existing in the formation.

It is well-known that the thickness of free product found in the formation (the actual thickness) is much smaller than that floating on top of the water in the groundwater monitoring well (the apparent thickness). Using the apparent thickness, without

FIGURE 4.6 A sample collected from a groundwater monitoring well at 110 ft bgs showing LNAPL floating on top of sediment-laden water. (Photo credit: Michael Shiang.)

any adjustment, to estimate the volume of free product may lead to an overestimate of the free product volume and an overdesign of the remediation system. The overestimate of free product in the RI phase may cause difficulties in obtaining an approval for final site closure because the remedial action can never recover the full amount of free product reported in the site assessment report.

Factors affecting the difference between the actual thickness and the apparent thickness include the densities (or specific gravity) of the free product and the characteristics of the formation (especially the pore sizes). Several approaches have been presented in the literature to correlate these two thicknesses. Ballestero et al. (1994) developed an equation using heterogeneous fluid flow mechanics and hydrostatics to determine the actual free product thickness in an unconfined aquifer. The equation is

$$t_g = t(1 - SG) - h_a \qquad (4.4)$$

where t_g = actual (formation) free product thickness; t = apparent (wellbore) product thickness; SG = specific gravity of the free product; and h_a = distance from the bottom of the free product to the water table (the free product above the capillary fringe is a distance from the water table). If no further data for h_a are available, average wetting capillary rise can be used as h_a.

Example 4.4: Determine the True Thickness of the Free-Floating Product

A recent survey of a groundwater monitoring well showed a 75-in. (190.5 cm) thick layer of gasoline floating on top of the water. The density of gasoline is 0.8 g/cm³ and the thickness of the capillary fringe above the water table is 1 ft (30.5 cm). Estimate the actual thickness of the free-floating product in the formation.

SOLUTION:

Using Eq. 4.4, the actual free product thickness in the formation is

$$t_g = (75 \text{ in.})(1-0.8) - 12 \text{ in.} = 3 \text{ in.} \ (7.6 \text{ cm})$$

Discussion:

1. Specific gravity, as described in Chapter 2, is the ratio of the density of a substance to the density of a reference substance (usually water at 4°C).
2. As shown in this example, the actual thickness of the free product is only 3 in. (7.6 cm), while the apparent thickness within the monitoring well is much larger, at 75 in. (190.5 cm), a 25-fold difference.

Example 4.5: Estimate the Mass and Volume of the Free-floating Product

Recent results from groundwater monitoring at an impacted site indicate that the areal extent of the free-floating product has an approximately rectangular shape of 50 ft×40 ft. From the apparent thicknesses of free product in four monitoring wells inside the plume, the true thicknesses of free product in the vicinities of these four wells were estimated to be 2, 2.6, 2.8, and 3 ft, respectively. The effective porosity of the subsurface is 0.35. Estimate the mass and volume of the free-floating product present at the site. Assume the specific gravity of the free-floating product is equal to 0.8.

SOLUTION:

(a) The areal extent of the free-floating product = (50 ft)(40 ft) = 2,000 ft²
(b) The average thickness of the free-floating product = (2+2.6+2.8+3)/4 = 2.6 ft
(c) The volume of the free-floating product
   = (volume of the free-floating product zone) × (effective porosity of the formation)
   = [(area)(thickness)] × (effective porosity of the formation)
   = [(2,000 ft²)(2.6 ft)](0.35) = (5,200 ft³)(0.35) = 1,820 ft³ = 13,610 gallons
(d) Mass of the free-floating product
   = (volume of the free-floating product)(density of the free-floating product)
   = (1,820 ft³)(0.8×62.4 lb/ft³)
   = 90,854 lb = 41,300 kg

Discussion: "Effective porosity" should be used instead of "porosity" for these type of estimates. The "effective porosity" represents the portion of pore space that is interconnected and contributes to the flow of the fluid (i.e., free product here) through the porous medium.

4.2   SOIL AND GROUNDWATER SAMPLING

Soil and groundwater sampling are an integral part of site assessment and remediation. Taking representative samples to obtain accurate information is critical for

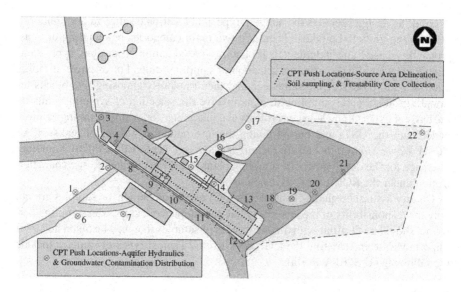

FIGURE 4.7 A map showing a site characterization plan of soil sampling (dotted lines) and groundwater sampling (circles). (Source: U.S. EPA 2005a.)

scientifically based decision making. A good soil sampling program consists of a well-developed sampling design, adequate sampling methods, and proper types and amounts of samples taken. Figure 4.7 is an example of a site plan where soil and groundwater samples are planned.

Other sampling techniques are geophysical techniques that can be used with field analytical and soil sampling equipment to better define the extent of subsurface contamination (U.S. EPA 1993a). Soil vapor surveys consist of placing sample tubes into shallow depths of the vadose zone. The soil gas is then pumped, collected, and analyzed. This is a good tool for locating VOC plumes in the vadose zone and for detecting VOC plumes in groundwater (U.S. EPA 1991a, 2020).

4.2.1 Sampling Design

Developing a sampling design is the first and crucial step to collecting appropriate and defensible data that accurately represent the problem being investigated. EPA's *Guidance on Choosing a Sampling Design for Environmental Data Collection for Use in Developing a Quality Assurance Project Plan* (U.S. EPA 2002a) is a good reference for developing a sound sampling program. To generate accurate information about the extent of contamination, the following should be considered:

- Appropriateness and accuracy of the sample collection and handling method
- Effect of measurement error
- Quality and appropriateness of the laboratory analysis
- Representativeness of the data with respect to the objective of the study

A sampling design specifies the number, type, and location (spatial and/or temporal) of sampling units to be taken. There are two main categories of sampling designs: probability-based and judgmental. Probability-based sampling designs apply sampling theory and involve random selection of sampling units. For example, dividing a lot into 100 blocks of the same size and then randomly choosing five blocks for sampling. Judgmental sampling designs involve the selection of sampling units on the basis of expert knowledge or professional judgment, for example, taking samples underneath the USTs and buried piping that are potential sources of leaks. EPA's *RCRA Waste Sampling – Draft Technical Guidance* (U.S. EPA 2002b) is another reference for the development of sampling plans, especially related to site characterization under the RCRA corrective action plans.

Quality assurance/quality control (QA/QC) requirements for the samples are not only the responsibility of the laboratory but the sample collectors. The sampling personnel should be familiar with different types of samples that may be taken and their importance for interpreting the analytical results. Typical types of samples include the following (U.S. EPA 1991a):

Field Samples

These are the samples collected in the field that are representative of the site conditions and analyzed in the laboratory for compounds and other constituents of interest.

Field Blank, Laboratory Blanks, Rinse/Cleaning Samples

A field blank is a sample of purified water (e.g., distilled or deionized) from the laboratory, brought to the field, poured into a sample container, closed, and then returned to the laboratory as a sample (along with the other collected samples). The level of contamination of the field blank is the zero analyte signal for determining the limit of detection. The laboratory blank is purified water used in the laboratory and is analyzed in the same manner as the field samples. Its function is to determine if contamination has occurred in the laboratory. A rinse or cleaning blank is a sample of the final rinse of a sampling device in the field before it is used for the next sampling location (e.g., a new soil borehole or another groundwater monitoring well). Its function is to determine if a sample may have been contaminated from materials taken in the previous sample.

Duplicate Samples, Replicate Samples, and Split Samples

Duplicate samples are commonly collected but not analyzed unless it is later determined that an additional analysis is necessary. For example, there are typically four brass tubes in a split-spoon sampler. Only the middle two core samples are kept, and they are duplicate samples. These two samples can be similar, but they are not identical. Replicate samples are the subsamples of the same sample and are considered identical samples. One common use of replicate samples is to label them differently to estimate the precision of the laboratory's analytical results. Split samples are replicate samples, but they are often analyzed by two different laboratories (e.g., one is administered by the regulatory agency).

Spike Samples

Some field samples may be split and spiked with a known concentration of a reference standard in the laboratory to allow for estimates of accuracy and detection of potential matrix interferences.

Many other considerations are needed to collect representative field samples, for example:

- Sampling frequency
- Sufficient sample size for samples to be representative and for analysis
- Proper materials for sample containers and devices: they should be inert (e.g., not to react with the samples; no potential leaching of compounds; no adsorption of COCs)
- Volatilization of VOCs during sampling and transportation (e.g., an ice chest with sufficient ice is needed to store the samples after sampling and before the lab receives and analyzes them)
- Bottles designed to eliminate headspace used for VOC samples
- A health and safety plan prepared before field work is started, and daily health and safety meetings
- A chain-of-custody form to track the samples from collection to analysis
- Sample holding time (i.e., the sample needs to be analyzed before the end of the holding time, which can range from hours to months, depending on the analyte)

4.2.2 SOIL SAMPLING

The U.S. EPA has published many guidelines with regard to soil sampling, including the following:

- *Site Characterization for Subsurface Remediation* (U.S. EPA 1991a)
- *Description and Sampling of Contaminated Soils: A Field Pocket Guide* (U.S. EPA 1991b)
- *Subsurface Characterization and Monitoring Techniques – A Desk Reference Guide (Volume I: Solids and Ground Water)* (U.S. EPA 1993b)
- *Subsurface Characterization and Monitoring Techniques – A Desk Reference Guide (Volume II: The Vadose Zone, Field Screening and Analytical Methods)* (U.S. EPA 1993c)
- *Use of Airborne, Surface, and Borehole Geophysical Techniques at Contaminated Sites – A Reference Guide* (U.S. EPA 1993a)
- *Soil Screening Guidance: User's Guide, 2nd ed.* (U.S. EPA 1996)
- *Accurately Determining Volatile Organic Compound (VOC) Concentrations in Soil and Solid Matrices* (U.S. EPA 2005b)
- *Soil Sampling* (U.S. EPA 2014)
- *Soil Gas Sampling* (U.S. EPA 2020)

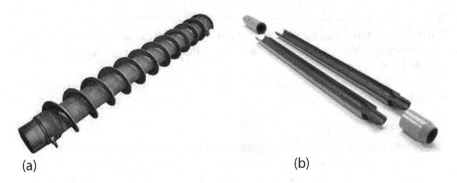

(a)   (b)

FIGURE 4.8   Photos of (a) a hollow-stem auger and (b) a split-spoon sampler.

Soil samples can be collected using a variety of methods and equipment, depending mainly on the depth of the desired sample (surface vs. subsurface), the type of sample required (disturbed vs. undisturbed, such as core samples), and the soil type (e.g., sandy vs. clayey). The collection of surface soil samples can be accomplished by using spades, spoons, scoops, and others. Sampling at greater depths may be performed using a hand or power auger (Ecology & Environment 1997). In shallow unconsolidated deposits, a hollow stem continuous flight auger is the preferred method to create a borehole/boring for soil sampling and for the installation of a groundwater well for monitoring or for water extraction (see Figure 4.8(a)). A split-spoon sampler is a thick wall tube split into two equal halves lengthwise. The two halves are locked together, typically containing four brass tubes, and the sampler is then driven into the undisturbed soil at the bottom of the borehole to take the soil samples (see Figure 4.8(b)).

The cuttings from soil borings are often temporarily stored on-site in 55-gallon (200 L) drums, such as those shown in Figure 4.9, before final disposal. It becomes necessary to estimate the volume of cuttings and the number of drums needed for storage. The calculation is relatively straightforward, as shown below.

To estimate the volume of cuttings from a soil boring, we calculate the volume of a cylinder and multiply by the fluffy factor, which takes into account the loosening of soil after removal from the subsurface:

$$\text{Volume of cuttings} = \Sigma\left(\frac{\pi}{4}d_b^2\right)(h)(\text{fluffy factor}) \qquad (4.5)$$

where d_b = diameter of the boring; and h = depth of the boring.

Example 4.6: Volume of Cuttings from a Soil Boring

Four 10-in. boreholes are to be drilled to 50 ft below the ground surface level for the installation of 4-in. groundwater monitoring wells. The fluffy factor is 1.2. Estimate the volume of soil cuttings and the number of 55-gallon drums needed to store the cuttings.

SOLUTION:

(a) Volume of cuttings from each boring $= [(\pi/4)(10/12)^2](50)(1.2) = 32.7$ ft^3
   Volumes of cutting from all four borings $= (4)(32.7) = 131$ ft^3

(b) Number of 55-gallon drums needed $= (131$ ft$^3)(7.48$ gallon/ft$^3) \div (55$ gallon/drum$) = 17.8$ drums
   Answer: Eighteen 55-gallon drums are needed.

4.2.3 GROUNDWATER SAMPLING

The EPA has published many guidelines with regard to groundwater sampling, including the following:

- *Compendium of ERT Groundwater Sampling Procedures* (U.S. EPA 1991c)
- *Handbook of Suggested Practices for the Design and Installation of Ground-Water Monitoring Wells* (U.S. EPA 1991d)
- *RCRA Ground-Water Monitoring – Draft Technical Guidance* (U.S. EPA 1992)
- *Subsurface Characterization and Monitoring Techniques – A Desk Reference Guide (Volume I: Solids and Ground Water)* (U.S. EPA 1993b)
- *Use of Airborne, Surface, and Borehole Geophysical Techniques at Contaminated Site – A Reference Guide* (U.S. EPA 1993a)
- *Ground-Water Sampling Guidelines for Superfund and RCRA Managers* (U.S. EPA 2002c)

FIGURE 4.9 55-gallon (200 L) drums used to store waste from a site, such as soil cuttings or contaminated water. (Photo credit: Michael Shiang.)

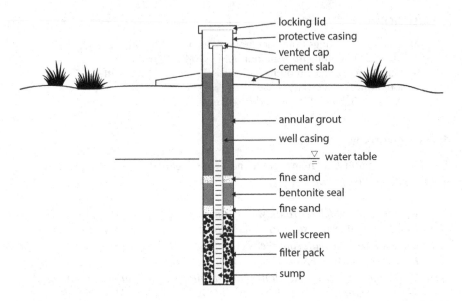

FIGURE 4.10   Components of a groundwater monitoring well. (Modified from U.S.G.S. 2017.)

Groundwater monitoring wells are installed to determine the characteristics of the aquifer (e.g., water level, water depth, and hydraulic conductivity) and to retrieve groundwater samples to determine the characteristics of the water in the aquifer. There are generally four types of monitoring activities: (1) detection monitoring to determine the presence of contamination, (2) assessment monitoring to determine the extent and magnitude of contamination, (3) evaluation monitoring to collect data for a remediation system design, and (4) performance monitoring to evaluate the effectiveness of the remediation effort.

A groundwater monitoring well consists of a well casing made of carbon steel, plastic, or stainless steel. It is installed into the ground within a drilled borehole or boring. The space between the casing and the borehole is called the annular space. The well has a permeable screen section at the bottom to allow groundwater to flow through. Inside the annular space, the well screen is surrounded by a filter pack to prevent fine soil grains from getting into the well. On top of the filter pack is usually a layer of bentonite, which has a large water-absorbing capacity to minimize the communication between the surface water and groundwater. The well casing is often secured in place by annular grout. The top of the casing is also secured (see Figure 4.10). Because often monitoring wells are constructed in traffic areas, the protective casing and lid are located below the ground surface, and a metal lid is installed flush with the pavement, as shown in the foreground in Figure 4.11.

Packing and seal materials need to be purchased and shipped to the site before the installation of the monitoring wells. A good estimate of the amount of packing material and bentonite seal is necessary for site assessment. To estimate the packing and

FIGURE 4.11 A monitoring well cover installed on a paved road. (Photo credit: Michael Shiang.)

seal materials needed, the following equation, which is the volume of the annular space between the well screen and the boring, can be used:

$$\text{Volume of packing materials or bentonite needed} = \frac{\pi}{4}\left(d_b^2 - d_c^2\right)(h) \quad (4.6)$$

where d_b = diameter of the boring, d_b = diameter of the well casing, and h = thickness interval of the well packing or bentonite seal. Note that this equation is the difference between two cylinders: the borehole and the well.

Example 4.7: Amount of Packing Materials Needed

The four monitoring wells in Example 4.6 are to be installed 15 ft into the ground-water aquifer. The wells are to be perforated with screens (0.02-in. slot opening) 15 ft below and 10 ft above the water table, with an additional 1 ft margin. Monterey Sand #3 is selected as the packing material. Estimate the number of 50-lb sand bags needed for this application. Assume the total bulk density of the sand is equal to 1.8 g/cm³ (112 lb/ft³).

SOLUTION:

(a) Packing interval for each well = perforation interval + 1 ft = (10 + 15) + 1 = 26 ft
Volume of sand needed for each well = {(π/4)[(10/12)² − (4/12)²]}(26) = 11.9 ft³
Volume of sand needed for four wells = (4)(11.9) = 47.6 ft³

(b) Number of 50-lb sand bags needed = (47.6 ft³)(112 lb/ft³) ÷ (50 lb/bag) = 107 bags
Answer: 107 bags are needed.

DISCUSSION:

1. The packing interval should be slightly larger than the perforation interval, so the 1 ft margin is added to part (a).
2. We should add an additional 10% to the estimate of sand usage as a safety factor to take into consideration that the borehole shape is not a perfect cylinder.

Example 4.8: Amount of Bentonite Seal Needed

The four monitoring wells in Examples 4.6 and 4.7 are to be sealed with 5 ft of bentonite below the top grout. Estimate the number of 50-lb bags of bentonite needed for this application. Assume the total bulk density of bentonite is equal to 1.8 g/cm³ (112 lb/ft³).

SOLUTION:

(a) Volume of bentonite needed for each well = {(π/4)[(10/12)² − (4/12)²]} (5) = 2.29 ft³
(b) Volume of bentonite needed for four wells = (2.29)(4) = 9.16 ft³
(c) Number of 50-lb bentonite bags needed = (9.16 ft³)(112 lb/ft³) ÷ (50 lb/ bag) = 20.5 bags

Answer: 21 bags are needed.

Discussion: We should add an additional 10% to the estimate of bentonite usage as a safety factor to take into consideration that borehole shape is not a perfect cylinder.

Groundwater samples can be retrieved using hand bailers, positive-displacement pumps, or low-flow submersible pumps. But stagnant water from a monitoring well must be removed before sampling, a process called purging. The stagnant volume includes the water inside the well casing and in the filter packing. A few parameters are often monitored, such as conductivity, pH, and temperature, to ensure they reach a consistent endpoint before sampling. The purge volume is site-specific and depends heavily on the subsurface geology. A guideline of purging three to five well volumes before groundwater sampling can be a starting point. The purged water is often impacted and needs to be treated, stored, and disposed of off-site. Figure 4.12 shows a monitoring well being purged, with its water passing through a pipe, hose, and flow meter setup, and being stored in a 55-gallon drum.

A good estimate of the volume of purged water is necessary for site assessment. To estimate the amount of purged water, the following equation can be used:

Well volume = Volume of the groundwater enclosed inside the well casing + Volume of the groundwater in the pore space of the packing

$$\text{Well volume} = \frac{\pi}{4} d_c^2(h) + \left[\frac{\pi}{4} \left(d_b^2 - d_c^2 \right)(h) \right] \phi \qquad (4.7)$$

where d_b = diameter of the boring, d_c = diameter of the well casing, h = depth of the well water, and ϕ = effective porosity of the packing.

FIGURE 4.12   A purging system for a groundwater monitoring well. (Photo credit: Michael Shiang.)

Example 4.9: Well Volume for Groundwater Sampling

The water depth inside one of the four monitoring wells in Examples 4.6–4.8 was measured to be 14.5 ft. The 4-in. diameter wells are installed inside 10-in. diameter boreholes. Three well volumes need to be purged before collecting a sample. Calculate the amount of purge water and the number of 55-gallon drums needed to store the water. Assume the effective porosity of the well packing is equal to 0.40.

SOLUTION:

(a) Well volume $= [(\pi/4) \times (4/12)^2 \times (14.5)] + \{(\pi/4) \times [(10/12)^2 - (4/12)^2] \times (14.5)\} \times (0.4) = 3.92$ ft^3

(b) Three well volumes $= (3)(3.92) = 11.8$ ft^3 for each well

(c) Number of 55-gallon drums needed for each well
$= [(11.8$ ft$^3)(7.48$ gallon/ft$^3)] \div (55$ gallon/drum$) = 1.6$ drums

(d) Total number of 55-gallon drums needed for four wells $= (1.6)(4) = 6.4$ drums

Answer: Seven 55-gallon drums are needed.

4.3   PLUME MIGRATION IN AN AQUIFER

Generally, from RI activities, the extent of the contaminated plume in subsurface soil and/or aquifer would be defined. If the chemicals of concern (COCs) are not removed, they may migrate further under common field conditions, and the plume(s) will enlarge.

In the vadose zone, the COCs will move downward as free product and, in the meantime, become dissolved in infiltrating water and then move downward by gravity. The downward-moving liquid may come into contact with the underlying aquifer and create a dissolved plume. The dissolved plume will move down-gradient in the aquifer (Figure 4.2). In addition, the COCs, especially VOCs, will volatilize into the air void of the vadose zone and travel under advective forces (with the air flow) and concentration gradients (through diffusion). Migration of the vapor can be in any direction, and the COCs in the vapor phase may come in contact with and get adsorbed by the soil moisture and groundwater. For site remediation or health risk assessment, understanding the fate and transport of COCs in the subsurface is important. Common questions related to the fate and transport of COCs in the subsurface include:

1. How long will it take for the plume in the vadose zone to enter the aquifer?
2. How far and how fast will the vapor COCs in the vadose zone travel? In what concentrations?
3. How fast does the groundwater flow? In which direction?
4. How fast will the plume migrate? In which direction?
5. Will the plume migrate at the same speed as the groundwater? If different, what are the factors that would make the plume migrate at a different speed?
6. How long has the plume been present in the aquifer?

This section and the next cover basic information needed to answer most of the above-mentioned questions. The coverage starts with groundwater movement and clarifies some common misconceptions about groundwater velocity and hydraulic conductivity. The discussion then moves to the migration of the plume in the aquifer and then in the vadose zone.

4.3.1 GROUNDWATER MOVEMENT

Darcy's law is commonly used to describe laminar flow in porous media. For a given medium, the flow rate is proportional to the head loss and inversely proportional to the length of the flow path. Flow in typical groundwater aquifers is laminar, and therefore Darcy's law is valid. Darcy's law can be expressed as

$$v_d = \frac{Q}{A} = K\frac{dh}{dl} \tag{4.8}$$

where v_d is the Darcy velocity, Q is the volumetric flow rate, A is the cross-sectional area of the porous medium perpendicular to the flow, dh/dl is the hydraulic gradient (a dimensionless quantity), and K is the hydraulic conductivity.

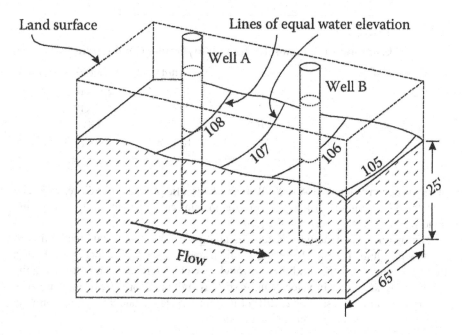

FIGURE 4.13 A perspective view of groundwater elevations and flow direction. (Source: Ward et al. 2014, © Taylor & Francis.)

Another common form of Darcy's law is

$$Q = K\frac{dh}{dl}A = KiA \qquad (4.9)$$

where i (= dh/dl) is the hydraulic gradient.

To determine the hydraulic gradient, dh/dl is calculated as the difference in groundwater heads divided by the length between the heads along the direction of groundwater flow:

$$i = \frac{dh}{dl} = \frac{h_1 - h_2}{L} \qquad (4.10)$$

where h_1 is the upgradient groundwater head, h_2 is the downgradient groundwater head directly along the direction of groundwater flow, and L is the total length between the two points.

Groundwater flow and elevations can be visualized using Figure 4.13, which is a perspective drawing of the subsurface and groundwater flow. Two wells, A and B, are installed into the subsurface. Through these wells, one can measure the groundwater elevation, shown as 108 ft in Well A and 106 ft in Well B. Groundwater flows downgradient from higher to lower elevation. In this drawing, groundwater is flowing

TABLE 4.2

Common Conversion Factors for Hydraulic Conductivity

m/d	cm/s	ft/d	gpd/ft^2
1	1.16×10^{-3}	3.28	2.45×10^1
8.64×10^2	1	2.83×10^3	2.12×10^4
3.05×10^{-1}	3.53×10^{-4}	1	7.48
4.1×10^{-2}	4.73×10^{-5}	1.34×10^{-1}	1

in the direction from Well A to Well B. The hydraulic gradient, therefore, would be calculated by taking the difference in groundwater elevations in the two wells and dividing by the length between the wells.

The hydraulic conductivity K tells us how permeable the porous medium is to the flowing fluid. It depends on the medium and the fluid. The larger the K of a formation, the easier the fluid flows through it. Commonly used units for hydraulic conductivity are either in velocity units such as ft/d, cm/s, or m/d, or in volumetric flow rate per unit area such as gpd/ft^2 or m^3/d/m^2. Table 4.2 tabulates some common conversion factors for hydraulic conductivity.

Example 4.10: Estimate the Rate of Groundwater Entering an Existing Plume

Leachates from a landfill leaked into the underlying aquifer and created a dissolved plume. Use the data below to estimate the amount of fresh groundwater that enters into the impacted zone per day:

- The maximum cross-sectional area of the plume perpendicular to the groundwater flow = 1,600 ft^2 (149 m^2), (20 ft in thickness × 80 ft in width = 6.1 m in thickness × 24.4 m in width)
- Hydraulic gradient = 0.005
- Hydraulic conductivity = 2,500 gpd/ft^2 (102 m^3/d/m^2)

SOLUTION:

The rate of fresh groundwater entering the plume can be found by inserting the appropriate values into Eq. 4.9:

$$Q = \left(2,500 \text{ gpd/ft}^2\right)(0.005)\left(1,600 \text{ ft}^2\right) = 20,000 \text{ gpd}$$

$$Q = \left(102 \text{ m}^3/\text{d/m}^2\right)(0.005)\left(149 \text{ m}^2\right) = 76 \text{ m}^3/\text{d}$$

DISCUSSION:

1. The calculation itself is straightforward. However, we can get valuable and useful information from this exercise. The rate of 20,000 gallons per

day (76 m³/d) represents the rate of upstream groundwater that will come into contact with the COCs. This water would become impacted and move downstream or side-stream and, consequently, enlarge the size of the plume.

2. To control the spread of the existing plume, one needs to extract this amount of water, 20,000 gpd (76 m³/d), or ~14 gallons/min (gpm) (53 L/min), at a minimum. The actual extraction rate required should be larger than this because the groundwater drawdown from pumping will increase the flow gradient. This increased gradient will, in turn, increase the rate of groundwater entering the impacted zone as indicated by the equation above. In addition, not all the extracted water will come from the impacted zone. More on this is discussed in Chapter 6.

3. Using the maximum cross-sectional area is a legitimate approach that represents the "contact face" between the fresh groundwater and the impacted zone. The maximum cross-sectional areas could be found as the product of the maximum plume thickness and the maximum plume width.

The velocity term in Eq. 4.8 is often called the Darcy velocity (or the discharge velocity). Does the Darcy velocity represent the actual groundwater flow velocity? The straight answer to this question is "no." The Darcy velocity in Eq. 4.8 assumes the flow occurs through the entire cross section of the porous medium. In other words, it is the velocity that water moves through an aquifer if the aquifer were an open, hollow, conduit. In reality, the flow is only through the available pore space. Therefore, the effective cross-sectional area available for flow is smaller than in a hollow conduit. Consequently, the actual fluid velocity through a porous medium would be larger than the corresponding Darcy velocity. This flow velocity is often called the seepage velocity or the interstitial velocity or the linear velocity. The relationship between the seepage velocity v_s and the Darcy velocity v_d is:

$$v_s = \frac{Q}{\phi A} = \frac{v_d}{\phi} \tag{4.11}$$

where ϕ is the effective porosity. For example, for an aquifer with an effective porosity of 33%, the seepage velocity will be three times the Darcy velocity (i.e., $v_s = 3\ v_d$).

Example 4.11: Estimate the Rate of Groundwater Entering the Existing Plume

An inert liquid spilled into the subsurface. The spill infiltrated the unsaturated zone and quickly reached the underlying water table aquifer. The aquifer consists mainly of sand and gravel with a hydraulic conductivity of 2,500 gpd/ft² (102 m³/d/m²) and an effective porosity of 0.35. The static water level in a well near the spill is 560 ft (171 m). The static water level in another well, 1 mile (1600 m) directly downgradient, is 550 ft (168 m). Determine the following:

- The Darcy velocity of the groundwater
- The seepage velocity of the groundwater

- The velocity of the plume migration
- How long it will take for the plume to reach the down-gradient well

SOLUTION:

(a) We need to determine the hydraulic gradient first:

$$USCS: i = dh/dl = \frac{(560 - 550)}{(5,280)} = 1.89 \times 10^{-3} \text{ft/ft} = 1.89 \times 10^{-3}$$

$$SI: i = dh/dl = \frac{(171 - 168)}{(1,600)} = 1.88 \times 10^{-3} \text{m/m} = 1.88 \times 10^{-3}$$

Darcy velocity $(v_d) = Ki$

$$USCS: v_d = \left[\left(2,500 \frac{gpd}{ft^2} \right) \left(0.134 \frac{\frac{ft}{d}}{\frac{gpd}{ft^2}} \right) \right] \left(1.89 \times 10^{-3} \right) = 0.63 \frac{ft}{d}$$

$$SI: v_d = \left(102 \frac{m}{d} \right) \left(1.88 \times 10^{-3} \right) = 0.19 \frac{m}{d}$$

(b) Seepage velocity $(v_s) = v_d / \phi$

$$USCS: v_s = 0.63 / 0.35 = 1.81 \text{ ft/d}$$

$$SI: v_s = 0.19 / 0.35 = 0.55 \text{ m/d}$$

(c) The pollutant is inert, meaning that it will not react with the aquifer materials (sodium chloride is a good example of an inert substance and is one of the common tracers used in aquifer studies). Therefore, the velocity of the plume for this case is the same as the seepage velocity, 1.81 ft/d (0.55 m/d).

$$USCS: Time = distance/velocity = (5,280 \text{ ft}) \div (1.81 \text{ ft/d}) = 2,900 \text{ days} = 8.0 \text{ years}$$

$$SI: Time = distance/velocity = (1,600 \text{ m}) \div (0.55 \text{ m/d}) = 2,900 \text{ days} = 8.0 \text{ years}$$

DISCUSSION:

1. The conversion factor (1 gpd/ft² = 0.134 ft/d), used in part (a), is from Table 4.2.
2. The calculated plume migration velocity is crude at best and should only be considered as a rough estimate. Many factors, such as hydrodynamic dispersion, are not considered in this equation. Dispersion can cause parcels of water to spread transversely to the main direction of groundwater flow. Dispersion is caused by factors such as tortuosity and intermixing of water particles due to differences in interstitial velocity induced by the heterogeneous pore sizes.

3. The migration speeds of most chemicals in a groundwater plume will be slowed down by interactions with aquifer materials, especially with clays, organic matter, and metal oxides and hydroxides. This phenomenon is called retardation and will be discussed further in the next section.

4.3.2 GROUNDWATER FLOW GRADIENT AND FLOW DIRECTION

Having a good knowledge of the gradient and direction of groundwater flow is vital to groundwater remediation. The gradient and direction of flow have great impacts on selection of remediation schemes to control plume migration such as location of the pumping wells and groundwater extraction rates, etc.

Estimates of the gradient and direction of groundwater flow can be made from a minimum of three monitoring wells with available measurements of groundwater elevations. The general procedure is described below, and an example follows.

Step 1: Locate the surveyed points on a map to scale.

Step 2: Mark the groundwater elevations next to the points on the map.

Step 3: Connect the points with a line.

Step 4: Subdivide each line into a number of segments of equal size (each segment represents an increment of groundwater elevation).

Step 5: Connect the points of equal values of elevation (equipotential lines), which then form the groundwater contours.

Step 6: Draw a line that passes through and is perpendicular to each equipotential line. This line marks direction of flow.

Step 7: Calculate the hydraulic gradient from the formula, $i = dh/dl$, where dh is the difference between the most upgradient and most downgradient equipotential lines and dl follows the direction of flow, which is not necessarily a straight line.

Example 4.12: Estimate the Gradient and Direction of Groundwater Flow from Three Groundwater Elevations

Three groundwater monitoring wells, A, B, and C, were installed at an impacted site. Groundwater elevations were determined from a recent survey of these wells and the values were marked on a map as shown in Figure 4.14. Estimate the hydraulic gradient and direction of the groundwater flow in the underlying aquifer.

SOLUTION:

(a) Water elevations (36.2 ft, 35.6 ft, and 35.4 ft) were measured at three monitoring wells and marked on the map in Figure 4.14.

(b) These three points are connected by straight lines to form a triangle.

(c) Subdivide each line of the triangle into a number of segments of equal intervals. For example, subdivide the line connecting point A (36.2 ft) and point B (35.6 ft) into three intervals. Each interval represents a 0.2 ft increment in elevation.

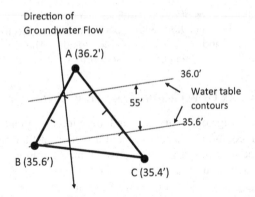

FIGURE 4.14   Determination of hydraulic gradient and direction (Example 4.12). (Source: Kuo 2014, © Taylor & Francis.)

(d) Connect the points of equal values of elevation (equipotential lines), which then form the groundwater contours. Here, we connect the elevations of 35.6 ft and 36.0 ft to form two contour lines.
(e) Draw a line that passes through and is perpendicular to each equipotential line and mark it as the groundwater flow direction.
(f) Measure the distance between two contour lines, 55 ft in this example shown in Figure 4.14.
(g) Calculate the hydraulic gradient from the formula, $i = dh/dl$:

$$i = (36.0 - 35.6)/(55) = 0.0073$$

Discussion: The groundwater elevations, especially those of the water table aquifers, may change with time. Consequently, the groundwater flow gradient and direction would change. Periodic surveys of the groundwater elevation may be necessary, if fluctuation of the water table is suspected. Off-site pumping, seasonal change, and recharge are some of the reasons that may cause the fluctuation of the water table elevation.

Example 4.12 demonstrates the determination of groundwater flow direction based on three wells. But many more wells can be used. Figure 4.15 is an example of a site with six monitoring wells, with groundwater contours (equipotential lines) ranging from 987.2 ft to 986.9 ft. To determine the hydraulic gradient of groundwater at the site, one calculates dh as the difference between the two extreme groundwater contours (987.2 – 986.9). The length dl is measured as the length of flow between those two contour lines, running perpendicular to the contour lines in between. In this case, $dl = 284$ ft, so $i = dh/dl = 0.0010$.

4.3.3   HYDRAULIC CONDUCTIVITY VS. INTRINSIC PERMEABILITY

In the soil venting literature, one may encounter a statement such as "the soil permeability is 4 darcys." In the groundwater remediation literature, one may read "the

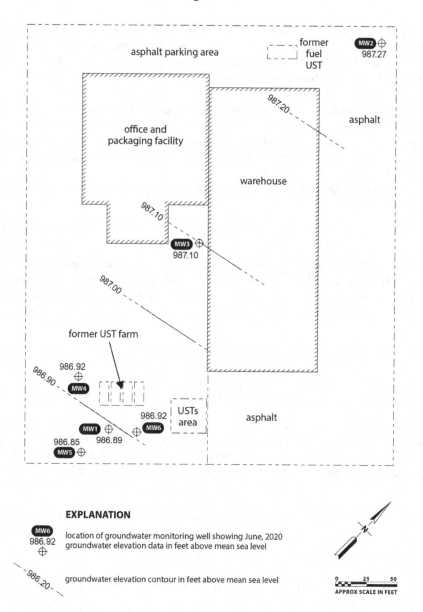

FIGURE 4.15 An industrial site with six monitoring wells used to determine groundwater equipotential lines.

hydraulic conductivity is equal to 0.05 cm/s." Both statements describe how permeable the formations are. Are they the same? If not, what is the relationship between the permeability and hydraulic conductivity?

These two terms, permeability and hydraulic conductivity, are sometimes used interchangeably. However, they have different meanings. The intrinsic permeability of a porous medium (e.g., subsurface soil or aquifer) defines its ability to transmit a fluid. It is a property of the medium only and is independent of the properties of the transmitting fluid. That is probably the reason why it is called the "intrinsic" permeability. On the other hand, the hydraulic conductivity of a porous medium depends on the properties of the fluid flowing through it and those of the medium itself.

Hydraulic conductivity is conveniently used to describe the ability of an aquifer to transmit groundwater. A porous medium has a unit hydraulic conductivity if it will transmit a unit volume of groundwater through a unit cross-sectional area (perpendicular to the direction of flow) in a unit time at the prevailing kinematic viscosity and under a unit hydraulic gradient.

The relationship between the intrinsic permeability and hydraulic conductivity is

$$K = \frac{k \rho g}{\mu}; \text{ or } k = \frac{K \mu}{\rho g} \tag{4.12}$$

where K is the hydraulic conductivity, k is the intrinsic permeability, μ is the fluid viscosity, ρ is the fluid density, and g is the gravitational constant (*Note*: kinematic viscosity $= \mu/\rho$, where μ is absolute viscosity and ρ is density). The intrinsic permeability has a unit of area as shown below:

$$k = \frac{K \mu}{\rho g} = \left[\frac{\left(\frac{m}{s}\right)\left(\frac{kg}{m \cdot s}\right)}{\left(\frac{kg}{m^3}\right)\left(\frac{m}{s^2}\right)} \right] = \left[m^2 \right] \tag{4.13}$$

In the petroleum industry, the intrinsic permeability of a formation is often expressed in the units of darcy. A formation has an intrinsic permeability of 1 darcy if it can transmit a flow of 1 cm³/s with a viscosity of 1 centipoise (1 mPa·s) under a pressure gradient of 1 atm/cm acting across an area of 1 cm² (*Note*: 1 Pa = 1 N/m²). That is,

$$1 \text{ darcy} = \frac{\left(1 \frac{cm^3}{s}\right)\left(10^{-3} Pa \cdot s\right)}{\left(1 \frac{atm}{cm}\right)\left(1 cm^2\right)} \tag{4.14}$$

By substituting appropriate units for atmosphere (i.e., 1 atm = 1.013×10^5 Pa), it can be shown that

$$1 \text{ darcy} = 9.87 \times 10^{-9} \text{ cm}^2 \tag{4.15}$$

Table 4.3 lists the mass density and viscosity of water under 1 atm. As shown in the table, the density of water from 0 to 40°C is essentially the same, at approximately 1 g/cm³; the viscosity of water decreases with increasing temperature. The viscosity of water at 20°C is 1 centipoise. (*Note*: This is the viscosity value of the fluid used in defining the darcy unit.)

TABLE 4.3

Physical Properties of Water under 1 atm

Temperature (°C)	Density (g/cm³)	Viscosity (cp)
0	0.999842	1.787
3.98	1.000000	1.567
5	0.999967	1.519
10	0.999703	1.307
15	0.999103	1.139
20	0.998207	1.002
25	0.997048	0.890
30	0.995650	0.798
40	0.992219	0.653

Note: 1 g/cm³ = 1,000 kg/m³ = 62.4 lb/ft³
1 centipoise (cp) = 0.01 poise = 0.01 g/cm·s = 0.001 Pa·s = 2.1 × 10⁻⁵ lb·s/ft²

Example 4.13: Determine Hydraulic Conductivity from a Given Intrinsic Permeability

The intrinsic permeability of a soil core sample is 1 darcy. What is the hydraulic conductivity of this soil for water at 15°C? How about at 25°C?

SOLUTION:

(a) At 15°C,
Density of water (15°C) = 0.999103 g/cm³ (from Table 4.3)
Viscosity of water (15°C) = 0.01139 poise = 0.01139 g/s·cm (from Table 4.3)

$$K = \frac{k\rho g}{\mu} = \frac{(9.87 \times 10^{-9}\,\text{cm}^2)\left(0.999103\,\frac{g}{cm^3}\right)\left(981\,\frac{cm}{s^2}\right)}{0.01139\,\frac{g}{s \cdot cm}} = 8.49 \times 10^{-4}\,\frac{cm}{s} = 18\,\frac{gpd}{ft^2}$$

(b) At 25°C,
Density of water (25°C) = 0.997048 g/cm³ (from Table 4.3)
Viscosity of water (25°C) = 0.00890 poise = 0.00890 g/s·cm (from Table 4.3)

$$K = \frac{k\rho g}{\mu} = \frac{(9.87 \times 10^{-9}\,\text{cm}^2)\left(0.997048\,\frac{g}{cm^3}\right)\left(981\,\frac{cm}{s^2}\right)}{0.00890\,\frac{g}{s \cdot cm}} = 1.08 \times 10^{-3}\,\frac{cm}{s} = 23\,\frac{gpd}{ft^2}$$

DISCUSSION:

1. A conversion factor (1 cm/s = 2.12 × 10⁴ gpd/ft²) from Table 4.2 was used.

2. As mentioned, hydraulic conductivity depends on the properties of the fluid flowing through it. This example illustrates that a porous medium with an intrinsic permeability of 1 darcy has a hydraulic conductivity of 18 gpd/ft² at 15°C and 23 gpd/ft² at 25°C. The hydraulic conductivity of this formation at a higher temperature (25°C) is larger than that at a lower temperature (15°C).
3. The intrinsic permeability is independent of temperature.
4. The unit of gpd/ft² is commonly used by hydrogeologists in the United States. The unit of cm/s is more commonly used in soil mechanics (e.g., the hydraulic conductivity of clay liners in landfills is commonly expressed in cm/s).

From the above example, one can tell that a geologic formation with an intrinsic permeability of 1 darcy has a hydraulic conductivity of approximately 10^{-3} cm/s or 20 gpd/ft² for transmitting pure water at 20°C. Typical values of intrinsic permeability and hydraulic conductivity for different types of formations are given in Table 4.4.

TABLE 4.4
Typical Values of Intrinsic Permeability and Hydraulic Conductivity

	Intrinsic Permeability	Hydraulic Conductivity	
	(darcy)	(cm/s)	(gpd/ft²)
Clay	10^{-6}–10^{-3}	10^{-9}–10^{-6}	10^{-5}–10^{-2}
Silt	10^{-3}–10^{-1}	10^{-6}–10^{-4}	10^{-2}–1
Silty sand	10^{-2}–1	10^{-5}–10^{-3}	10^{-1}–10
Sand	1–10^2	10^{-3}–10^{-1}	10–10^3
Gravel	10–10^3	10^{-2}–1	10^2–10^4

4.4 MIGRATION VELOCITY OF THE DISSOLVED PLUME

As COC spills enter the subsurface, the materials may move downward as free-product or be dissolved into the infiltrating water and then move downward by gravity. This liquid may travel deep enough to get in contact with the underlying aquifer and form a dissolved plume in the aquifer. This section will discuss the migration of the dissolved plume, which is relatively simpler than the transport of the COCs in the vadose zone. This discussion is applicable to many types of COCs, such as VOCs, semi-VOCs, and heavy metals. Transport in the vadose zone will be discussed in Section 4.5.

4.4.1 THE ADVECTION-DISPERSION EQUATION

The design and selection of optimal remediation schemes, such as the number and locations of extraction wells, often require the prediction of the COC distribution in the subsurface over time. These predictions are then used to evaluate different remediation scenarios. To make such predictions, we need to couple the equation describing the flow with the equation of mass balance.

To describe the fate and transport of a COC, the one-dimensional form of the advection-dispersion equation can be expressed as:

$$\frac{\partial C}{\partial t} = D\frac{\partial^2 C}{\partial x^2} - v\frac{\partial C}{\partial x} \pm RXNs \qquad (4.16)$$

where C is the COC concentration, D is the dispersion coefficient, v is the velocity of the fluid flow, t is the time, and RXNs represents the reactions. Eq. 4.16 is a general equation and is applicable to describe the fate and transport of COCs in the vadose zone or in the groundwater. The first term in Eq. 4.16 describes the change in COC concentration in the fluid, contained within a specific volume of an aquifer or a vadose zone, with time. The first term on the right-hand side describes the net dispersive flux of the COC in and out of the fluid in this volume, and the second term describes the net advective flux. The last term represents the amount of COC that may be added or lost to the fluid in this volume due to physical, chemical, and/ or biological reactions. For plume migration in groundwater, v is the groundwater seepage velocity that can be determined from the Darcy's law and the porosity of the aquifer (i.e., Eq. 4.11).

4.4.2 DIFFUSIVITY AND DISPERSION COEFFICIENT

The dispersion term in Eq. 4.16 accounts for both the molecular diffusion and hydraulic dispersion. The molecular diffusion, strictly speaking, is due to the concentration gradient (i.e., the concentration difference). The COC diffuses away from the higher concentration zone, and this can occur even when the fluid is not moving. The hydraulic dispersion here is mainly caused by flow in porous media. It results from (i) the velocity variation within a pore, (ii) the different pore geometries, (iii) the divergence of flow lines around the soil grains in the porous media, and (iv) the aquifer heterogeneity (U.S. EPA 1989).

The dimension of the dispersion coefficient is $(length)^2/(time)$. Field studies of the dispersion coefficient revealed that it varies with groundwater velocity. They show that the dispersion coefficient is relatively constant at low velocities (where the molecular diffusion dominates the COC spread) but increases linearly with velocity as the groundwater velocity increases (when the hydraulic dispersion dominates). The dispersion coefficient can be written as the sum of two terms: effective molecular diffusion coefficient D_d and hydraulic dispersion coefficient D_h:

$$D = D_d + D_h \qquad (4.17)$$

The effective molecular diffusion coefficient can be obtained from the molecular diffusion coefficient D_o as:

$$D_d = (\xi)(D_o) \qquad (4.18)$$

where ξ is the tortuosity factor that accounts for the increased distance that COCs need to travel to get around the soil grains. Typical ξ values are in the range of 0.6–0.7 (EPA 1989).

The hydraulic dispersion coefficient is proportional to the groundwater flow velocity as

$$D_h = (\alpha)(v) \tag{4.19}$$

where α is the dispersivity. The hydraulic dispersion coefficient is scale-dependent; its value has been observed to increase with increasing transport distance. The longitudinal dispersivity values from field tracer tests and model calibration of COC plumes are found to be in the range of 10–100 m, which is much higher than that from column studies in the laboratories.

The molecular diffusion coefficients of COCs in dilute aqueous solutions are much smaller than in gases at atmospheric pressure and usually range from 0.5×10^{-5} to 2×10^{-5} cm²/s at 25°C (compared to typical values of 0.05 to 0.5 cm²/s in the gaseous phase, as shown in Table 2.2). Values of molecular diffusion coefficients of selected compounds are shown in Table 4.5.

The diffusion coefficient of a compound can be estimated from using the diffusion coefficient of another compound of similar species, their molecular weights, and the following relationship:

$$\frac{D_1}{D_2} = \sqrt{\frac{MW_2}{MW_1}} \tag{4.20}$$

As shown in Eq. 4.20, the diffusion coefficient is inversely proportional to the square root of its molecular weight. The heavier the COC, the harder it is for it to diffuse through the fluid. Temperature also has an influence on the diffusion coefficient.

TABLE 4.5

Values of Diffusion Coefficients of Selected Compounds in Water

Compound	Temperature (°C)	Diffusion Coefficient (cm²/s)
Acetone	25	1.28×10^{-5}
Acetonitrile	15	1.26×10^{-5}
Benzene	20	1.02×10^{-5}
Benzoic acid	25	1.00×10^{-5}
Butanol	15	0.77×10^{-5}
Ethylene glycol	25	1.16×10^{-5}
Propanol	15	0.87×10^{-5}

Source: Sherwood et al. 1975.

The diffusion coefficient in water is proportional to the temperature and inversely proportional to the fluid viscosity. The water viscosity (μ_w) decreases with increasing temperature, and, consequently, the diffusion coefficient increases with temperature, and the following relationship applies:

$$\frac{D @ T_1}{D @ T_2} = \left(\frac{T_1}{T_2}\right)\left(\frac{\mu_w @ T_2}{\mu_w @ T_1}\right) \tag{4.21}$$

Example 4.14: Estimate the Diffusion Coefficient at Different Temperatures

The diffusion coefficient of benzene in a dilute aqueous solution at 20°C is 1.02×10^{-5} cm²/s (Table 4.5). Use this reported value to estimate:

- The diffusion coefficient of toluene in a dilute aqueous solution at 20°C
- The diffusion coefficient of benzene in a dilute aqueous solution at 25°C

SOLUTION:

(a) The MW of toluene ($C_6H_5CH_3$) is 92 g/mol, and the MW of benzene (C_6H_6) is 78 g/mol.
Using Eq. 4.20:

$$\frac{D_1}{D_2} = \frac{\left(1.02 \times 10^{-5}\right)}{D_2} = \sqrt{\frac{92}{78}}$$

So the diffusion coefficient of toluene at 20°C = 0.94×10^{-5} cm²/s.

(b) Viscosity of water at 20°C = 1.002 cp (from Table 4.4)

Viscosity of water at 25°C = 0.89 cp (from Table 4.4)
Using Eq. 4.21:

$$\frac{\left(1.02 \times 10^{-5}\right)}{D @ 298K} = \left(\frac{293}{298}\right)\left(\frac{0.89}{1.002}\right)$$

So the diffusion coefficient of benzene at 25°C = 1.17×10^{-5} cm²/s.

Discussion: The diffusion coefficient of benzene at 25°C is about 15% larger than at 20°C.

Example 4.15: Relative Importance of Molecular Diffusion and Hydraulic Dispersion

Benzene from leaky USTs at a site leaked into the underlying aquifer. The hydraulic conductivity of the aquifer is 500 gpd/ft² and the effective porosity is 0.4. The groundwater temperature is 20°C. The dispersivity is found to be 2 m. The tortuosity factor is 0.65. Estimate the relative importance between the hydraulic dispersion and the molecular diffusion for the dispersion of the benzene plume in the following two cases: (a) the hydraulic gradient=0.01; (b) the hydraulic gradient=0.0005.

SOLUTION:

(a) The hydraulic conductivity of the aquifer=500 gpd/ft^2
= (500)(4.73$\times$10^{-5})=0.024 cm/s (Use the conversion factor in Table 4.2)
Use Eqs. 4.9 and 4.11 to find the groundwater velocity (for gradient=0.01).

$$V_s = \frac{(0.024)(0.01)}{0.4} = 6.0 \times 10^{-4} \text{ cm/s}$$

The molecular diffusion coefficient of benzene (at 20°C)=1.02$\times$10^{-5} cm^2/s (Table 4.5).
The effective molecular diffusion coefficient can be obtained as (Eq. 4.18):

$$D_d = \xi(D_o) = (0.65)(1.02 \times 10^{-5}) = 0.66 \times 10^{-5} \text{ cm}^2/\text{s}$$

The hydraulic dispersion coefficient can be determined as (Eq. 4.19):

$$D_h = \alpha(v) = (200 \text{ cm})(6 \times 10^{-4} \text{ cm/s}) = 12,000 \times 10^{-5} \text{ cm}^2/\text{s}$$

The hydraulic dispersion coefficient is much larger than the diffusion coefficient. Therefore, the hydraulic dispersion will be the dominant mechanism for the dispersion of COCs.

(b) For a smaller gradient, the groundwater will move more slowly, and the dispersion coefficient will be proportionally smaller. The effective molecular diffusion coefficient will be the same as in part (a), 0.66$\times$10^{-5} cm^2/s.

Use Eqs. 4.9 and 4.11 to find the groundwater velocity (for gradient=0.005):

$$V_s = \frac{(0.024)(0.0005)}{0.4} = 3.0 \times 10^{-5} \text{ cm/s}$$

The hydraulic dispersion coefficient can then be determined as (Eq. 4.19):

$$D_h = \alpha(v) = (200 \text{ cm})(3.0 \times 10^{-5} \text{ cm/s}) = 600 \times 10^{-5} \text{ cm}^2/\text{s}$$

The hydraulic dispersion coefficient is still much larger than the diffusion coefficient at this relatively flat gradient of 0.0005.

Discussion: In the second case, the groundwater movement is very slow at 3.0$\times$10^{-5} cm/s (or 31 ft/year), and the hydraulic dispersion is still the dominant mechanism (for dispersivity=2 m). The diffusion coefficient will become more important only if the flow rate and/or the dispersivity is smaller. Nonetheless, the molecular diffusion accounts for a common phenomenon where the plume usually extends slightly *upstream* of the entry point into the aquifer.

4.4.3 RETARDATION FOR PLUME MIGRATION IN GROUNDWATER

Many physical, chemical, and biological processes in the subsurface can affect the fate and transport of COCs, including biotic degradation, abiotic degradation, dissolution, ionization, volatilization, and adsorption. For the transport of a dissolved plume in groundwater, adsorption of COCs is probably the most important and most studied mechanism that removes COCs from the groundwater (adsorption was also discussed in Section 2.5.5). If adsorption is the primary removal mechanism in the subsurface, the reaction term in Eq. 4.16 can then be written as $(\rho_b/\phi)\partial S/\partial t$, where ρ_b is the dry bulk density of soil (or the aquifer matrix), ϕ is the porosity, t is time, and S is the COC concentration adsorbed onto the aquifer solids.

When the COC concentration is low, a linear adsorption isotherm is usually valid. Assume a linear adsorption isotherm (e.g., $S = K_p C$), thus

$$\frac{\partial S}{\partial C} = K_p \tag{4.22}$$

where K_p is the liquid-soil partition coefficient as seen in Chapter 2.

The following relationship can then be derived:

$$\frac{\partial S}{\partial C} = \left(\frac{\partial S}{\partial C}\right)\left(\frac{\partial C}{\partial t}\right) = K_p \frac{\partial C}{\partial t} \tag{4.23}$$

Substitute Eq. 4.23 into Eq. 4.16 and rearrange:

$$\frac{\partial C}{\partial t} + \left(\frac{\rho_b}{\phi}\right)K_p\frac{\partial C}{\partial t} = \left(1 + \frac{\rho_b K_p}{\phi}\right)\frac{\partial C}{\partial t} = D\frac{\partial^2 C}{\partial x^2} - v\frac{\partial C}{\partial x} \tag{4.24}$$

Divide both sides by $(1 + \rho_b K_p/\phi)$, and Eq. 4.23 can be simplified into the following form:

$$\frac{\partial C}{\partial t} = \frac{D}{R}\frac{\partial^2 C}{\partial x^2} - \frac{v}{R}\frac{\partial C}{\partial x} \tag{4.25}$$

where

$$R = 1 + \frac{\rho_b K_p}{\phi} \tag{4.26}$$

The parameter R is the retardation factor (dimensionless) and has a value ≥ 1. Eq. 4.25 is essentially the same as Eq. 4.16 except that the reaction term in Eq. 4.25 is taken care of by R (Eq. 4.26). The retardation factor reduces the impact of dispersion and migration velocity by a factor of R. All of the mathematical solutions that are used to solve for the transport of inert tracers can be used for the transport of the COCs if the groundwater velocity and the dispersion coefficient are divided by the retardation factor. From the definition of R, we can tell that R is a function of ρ_b, ϕ, and K_p. For a given aquifer, ρ_b and ϕ would be the same for different COCs. Consequently, the larger the partition coefficient, the larger the retardation factor.

Example 4.16: Determination of the Retardation Factor

The aquifer underneath a site is impacted by several organic compounds, including benzene, 1,2-dichloroethane (DCA), and pyrene. Estimate their retardation factors using the following data from the site assessment:

- Effective aquifer porosity $=0.40$
- Dry bulk density of the aquifer materials $=1.6$ g/cm^3
- Fraction of organic carbon of the aquifer materials $=0.015$
- $K_{oc} = 0.63 \, K_{ow}$

SOLUTION:

(a) From Table 2.2,
   $Log(K_{ow}) = 2.13$ for benzene $\rightarrow K_{ow} = 135$ L/kg
   $Log(K_{ow}) = 1.53$ for 1,2-DCA $\rightarrow K_{ow} = 34$ L/kg
   $Log(K_{ow}) = 4.88$ for pyrene $\rightarrow K_{ow} = 75,900$ L/kg

(b) Using Eq. 2.23, $K_{oc} = 0.63 K_{ow}$, from Section 2.5.5, we obtain:
   $K_{oc} = (0.63)(135) = 85$ L/kg (benzene)
   $K_{oc} = (0.63)(34) = 22$ L/kg (1,2-DCA)
   $K_{oc} = (0.63)(75,900) = 47,800$ L/kg (pyrene)

(c) Using Eq. 2.21 from Section 2.5.5, $K_p = f_{oc} K_{oc}$, and $f_{oc} = 0.015$, we obtain:
   $K_p = (0.015)(85) = 1.275$ L/kg (benzene)
   $K_p = (0.015)(22) = 0.32$ L/kg (1,2-DCA)
   $K_p = (0.015)(47,800) = 717$ L/kg (pyrene)

(d) Use Eq. 4.26 to find the retardation factor:

$$R = 1 + \frac{\rho_b K_p}{\phi} = 1 + \frac{(1.8)(1.275)}{0.4} = 6.10 \quad \text{for benzene}$$

$$R = 1 + \frac{\rho_b K_p}{\phi} = 1 + \frac{(1.6)(0.32)}{0.4} = 2.28 \quad \text{for 1,2-DCA}$$

$$R = 1 + \frac{\rho_b K_p}{\phi} = 1 + \frac{(1.6)(717)}{0.4} = 2,869 \quad \text{for pyrene}$$

DISCUSSION:

1. Pyrene is very hydrophobic with a large K_p value, and its retardation factor is much larger than those of benzene and 1,2-DCA.
2. The units of K_{ow}, K_{oc}, and K_p are typically L/kg, which is equivalent to mL/g. When used in Eq. 4.26, mL/g cancels out with the dry bulk density units of g/cm^3 because 1 mL = 1 cm^3.

4.4.4 MIGRATION OF THE DISSOLVED PLUME

The retardation factor relates the plume migration velocity to the groundwater seepage velocity as

$$R = \frac{v_s}{v_p} \qquad (4.27)$$

where v_s is the groundwater seepage velocity and v_p is the velocity of the dissolved plume. When the value of R is equal to unity (for inert compounds), the compound will move at the same speed as the groundwater flow without any "retardation," or slowing down. When $R = 2$, for example, the COC will move at half of the ground-water flow velocity.

Example 4.17: Migration Speed of Dissolved Plume in Groundwater

The aquifer underneath a site is impacted by several organic compounds including benzene, 1,2-dichloroethane (DCA), and pyrene. A recent groundwater monitor-ing event in September 2020 indicated that 1,2-DCA and benzene have traveled 250 m and 20 m downgradient, respectively, while no pyrene compounds were detected in the downgradient wells. Estimate the time when the COCs first entered the aquifer. The following data were obtained from the site assessment:

- Effective aquifer porosity = 0.40
- Aquifer hydraulic conductivity = 30 m/day
- Hydraulic gradient = 0.005
- Dry bulk density of aquifer materials = 1.6 g/cm^3
- Fraction of organic carbon of the aquifer materials = 0.015
- $K_{oc} = 0.63\ K_{ow}$

Briefly discuss your results and list possible factors that may cause your estimate to be different from the true value.

SOLUTION:

(a) Use Eq. 4.8 to find the Darcy velocity: $v_d = ki = (30)(0.005) = 0.15$ m/d
(b) Use Eq. 4.10 to find the groundwater velocity (i.e., the seepage velocity, or the interstitial or linear velocity): $v_s = v_d/\phi = (0.15)/(0.4) = 0.375$ m/d
(c) Use Eq. 4.27 and the values of R from Example 4.16 to determine the migration speeds of the plumes:
$v_p = (0.375)/(6.10) = 0.061$ m/d = 22.4 m/year (benzene)
$v_p = (0.375)/(2.28) = 0.164$ m/d = 60.0 m/year (1,2-DCA)
$v_p = (0.375)/(2,864) = 0.000131$ m/d = 0.048 m/year (pyrene)
(d) The time for 1,2-DCA to travel 250 m can be found as:
t = (distance)/(migration speed)
= (250 m)/(60.0 m/year) = 4.17 year = 4 years and 2 months
So 1,2-DCA entered the aquifer in July of 2016.
(e) The time for benzene to travel 50 m can be found as:
t = (50 m)/(22.4 m/year) = 2.23 year = 2 years and 3 months
So benzene entered the aquifer in June of 2018.

DISCUSSION:

1. The estimates are the times when benzene and 1,2-DCA first entered the aquifer. The data given are insufficient to estimate the time the leachates

traveled through the vadose zone and, consequently, the time the leaks started from the sources (e.g., leaky USTs).

2. The retardation factor of 1,2-DCA is smaller; therefore, its migration speed in the vadose zone would be faster. This helps to explain the fact that 1,2-DCA entered the aquifer earlier than benzene.

3. The migration speed of pyrene is extremely small, 0.042 m/year; therefore, it was not detected in the downstream monitoring wells. Most, if not all, of the pyrene compounds will be adsorbed onto the soil in the vadose zone. The pyrene may travel in the aquifer by adsorbing onto the colloidal particles.

4. The estimates are crude because lots of factors may affect the accuracy of the estimates. Factors include uncertainty in the values of the hydraulic conductivity, porosity, hydraulic gradient, K_{ow}, f_{oc}, etc. Neighborhood pumping will affect the hydraulic gradient and, consequently, the migration of the plume. Other subsurface reactions such as oxidation and biodegradation may also have large impacts on the fate and transport of these COCs.

4.5 RETARDATION FACTOR FOR COC VAPOR MIGRATION IN THE VADOSE ZONE

For an air stream flowing through a porous medium, the gas-phase retardation factor (R_a) can be derived as (U.S. EPA 1991a),

$$R_a = 1 + \frac{\rho_b K_p}{\phi_a H^*} + \frac{\phi_w}{\phi_a H^*} \tag{4.28}$$

where ρ_b is the dry bulk density of the soil, K_p is the soil-water partition coefficient, H^* is the dimensionless Henry's constant, ϕ_a is the air-filled porosity, and ϕ_w is the volumetric water content.

This gas-phase retardation factor will be a constant if ϕ_w does not change. It is analogous to the retardation factor R for the movement of COCs in an aquifer. The movement of the COC in the void of the vadose zone will be retarded by a factor of R_a. The second term on the right-hand side of Eq. 4.28 represents the partitioning of the COCs between the vapor phase, the soil moisture phase, and the solid phase. The third term represents the partitioning between the vapor phase and the soil moisture. As the COC in the vapor phase moves through the air-filled pores, the migration rate of the COC in the air is slower than that of the air itself because of the loss of its mass to the soil moisture and to the soil's organic carbon.

Under the condition of no advective flow, the gas-phase retardation factor can be defined as the ratio of the diffusion rate of an inert compound, such as nitrogen, to the diffusion rate of the COC. Under advective flow, it can be used as the relative measure to compare the migration rates of compounds with different retardation factors. For a soil vapor extraction application (covered in Chapter 5), the air-phase retardation factor is also the minimum number of pore volumes that must pass through the impacted zone to clean it up. It is considered as the minimum because this approach ignores the effects of mass transfer limitations among the phases, subsurface heterogeneity, and unequal travel time from the outer edge of the plume to the vapor extraction well (U.S. EPA 1991a).

As shown in Eq. 4.28, the air-phase retardation factor increases with ϕ_w and K_p but decreases with Henry's constant. A higher moisture content means a larger water reservoir to retain the COCs, and a larger K_p value indicates that the soil has a larger organic content or the COC is more hydrophobic. On the other hand, a compound with a larger Henry's constant would have a stronger tendency to volatilize into the air void. The Henry's constant increases with increasing temperature and, thus, a smaller air-phase retardation factor at a higher temperature. Therefore, for a soil vapor extraction application, at higher temperatures, fewer pore volumes of air need to be moved through the impacted zone to remove the COCs.

Example 4.18: Determination of the Air-Phase Retardation Factor

The vadose zone underneath a site is impacted by several organic compounds, including benzene, 1,2-dichloroethane (DCA), and pyrene. Estimate the air-phase retardation factor using the following data from the site assessment:

- Vadose zone soil porosity = 0.40
- Volumetric water content = 0.15
- Dry bulk density of soil = 1.6 g/cm³
- Fraction of soil organic carbon = 0.015
- Temperature of the formation = 25°C
- $K_{oc} = 0.63\ K_{ow}$

SOLUTION:

(a) From Table 2.2, $H = 5.55$ atm/M for benzene (at 25°C)
Use Table 2.3 to convert it to a dimensionless value:

$$H^* = H/RT = (5.55)/[(0.08206)(298)] = 0.227$$

Similarly, for 1,2-DCA (Henry's constant value in the table is for 20°C. We use this value for 25°C as an approximate value) and pyrene:

$$H^* = H/RT = (0.98)/[(0.08206)(298)] = 0.04\ \text{(for 1,2-DCA)}$$

$$H^* = H/RT = (0.005)/[(0.08206)(298)] = 0.0002\ \text{(for pyrene)}$$

(b) From Example 4.16,

$$K_p = (0.015)(85) = 1.275\ \text{L/kg}\quad \text{(for benzene)}$$

$$K_p = (0.015)(22) = 0.32\ \text{L/kg}\quad \text{(for 1,2-DCA)}$$

$$K_p = (0.015)(47,800) = 717\ \text{L/kg}\quad \text{(for pyrene)}$$

(c) Use Eq. 4.28 to find the air-phase retardation factor:

$$R_a = 1 + \frac{\rho_b K_p}{\phi_a H^*} + \frac{\phi_w}{\phi_a H^*} = 1 + \frac{(1.6)(1.275)}{(0.25)(0.227)} + \frac{(0.15)}{(0.25)(0.227)} = 39.6 \quad \text{for benzene}$$

$$R_a = 1 + \frac{\rho_b K_p}{\phi_a H^*} + \frac{\phi_w}{\phi_a H^*} = 1 + \frac{(1.6)(0.32)}{(0.25)(0.04)} + \frac{(0.15)}{(0.25)(0.04)} = 67.2 \quad \text{for 1,2-DCA}$$

$$R_a = 1 + \frac{\rho_b K_p}{\phi_a H^*} + \frac{\phi_w}{\phi_a H^*} = 1 + \frac{(1.6)(717)}{(0.25)(0.0002)} + \frac{(0.15)}{(0.25)(0.0002)} = 2.3 \times 10^7 \quad \text{for pyrene}$$

Discussion: Pyrene is very hydrophobic and has a low Henry's constant. Its air-phase retardation factor is much larger than those of benzene and 1,2-DCA.

4.6   SUMMARY

This chapter describes the activities required to investigate subsurface contamination before a remedial action plan can be designed:

- Determination of the amount of COC in the vadose zone and capillary fringe
- Collection of soil and groundwater samples
- Calculating the rate and direction of groundwater flow
- Determining the velocity of the COCs in a groundwater plume taking into account advection, diffusion, dispersion, and retardation
- Determining the retardation factor of COCs in the vadose zone

4.7   PROBLEMS AND ACTIVITIES

4.1.   After leaky USTs were removed from a site, soil borings were installed. Soil samples were taken every 2 m below ground surface. Based on the analyzed samples, the areas of the plume at a few depths were determined as follows. The total bulk density of soil is 1,700 kg/m³. Determine the volume and mass of the impacted soil in the vadose zone.

Depth (meter bgs)	Area of the plume at that depth (m²)
4	0
6	28
8	38
10	59
12	75
14	0

4.2. After leaky USTs were removed from a site, three soil borings were installed. Soil samples were taken every meter below ground surface. However, not all the samples were analyzed due to budget constraints. Based on the analyzed samples, the areas of the plume at a few depths were determined as shown below. The total bulk density of soil is 1,800 kg/m³. Determine the volume and mass of the impacted soil left in the vadose zone.

Depth (meter bgs)	Area of the plume at that depth (m²)
3	0
4	12
6	27
7	35
8	0

4.3. A recent survey of a groundwater monitoring well showed a 0.9-m layer of gasoline floating on top of the groundwater in the well. Use the following information to answer the questions below:
   - Density of the gasoline $= 0.75$ g/cm³
   - Pore radius of the capillary fringe (average) $= 0.20$ mm
   - Effective porosity of formation outside the well $= 35\%$
   - Area of the free-product (assuming the thickness is uniform) $= 200$ m²
   (a) Estimate the height of the capillary fringe, in meters.
   (b) Estimate the actual thickness of the free-floating product in the formation, in meters.
   (c) Estimate the amount of the free-product present in the formation, in m³ and gallons.

4.4. Four 10-in. boreholes were drilled to 75 ft below ground surface for installation of four groundwater monitoring wells. Assuming a fluffy factor of 1.1, estimate
   (a) the volume of soil cuttings from these four borings, in ft³.
   (b) the number of 55-gallon drums needed to store the cuttings from these four borings.

4.5. Five 10-in. boreholes were drilled to 50 ft below ground surface for installation of 4-in. groundwater monitoring wells into an unconfined aquifer. Monterey Sand #3 ($\rho_t = 1.8$ g/cm³ $= 112$ lb/ft³) was used for well packing, and the porosity of the packing was determined to be 0.4. During the most recent quarterly monitoring event, the average water depth inside these five wells was found to be 20 ft. Four well volumes of groundwater were purged out of each well before sampling.
   (a) Estimate the volume of water in one well, in gallons.
   (b) Estimate the total volume of water needed to be purged from these five wells before sampling, in gallons.

4.6. An industrial site used 1,2-dichloropropane ($C_3H_6Cl_2$) as a solvent since the late 1990s. Recently, a site assessment was conducted and found that the site has been impacted. The dissolved 1,2-dichloropropane plume in the underlying aquifer has traveled 100 m downstream at a speed of 40 m/year. With the following information, answer the questions below:

- Aquifer porosity = 0.4
- Hydraulic gradient = 0.02
- Bulk density of aquifer materials = 1.8 g/cm^3
- f_{oc} of aquifer materials = 0.01
- Log(K_{ow}) of 1,2-dichloropropane = 2.0

(a) Estimate the retardation factor for the plume transport.

(b) Estimate the hydraulic conductivity of the aquifer, in cm/s.

(c) The log(K_{ow}) of xylene is 3.0. If xylene is also present in the same aquifer, will it travel faster or slower than 1,2-dichloropropane?

(d) Provide the reason to your answer for (c).

REFERENCES

Ballestero, T. P., Fiedler, F. R., and Kinner, N. E. (1994). "An Investigation of the Relationship Between Actual and Apparent Gasoline Thickness in a Uniform Sand Aquifer." *Groundwater*, 32(5), 708.

Ecology & Environment. (1997). *Soil Sampling - U.S. EPA Standard Operating Procedure ENV 3.13*. Ecology & Environment, Inc., Lancaster, NY.

Fetter, Jr., C. W. (1980). *Applied Hydrogeology*. Charles E. Merrill Publishing, Columbus, OH.

Kuo, J. (2014). *Practical Design Calculations for Groundwater and Soil Remediation*. CRC Press, Boca Raton, Florida.

Sherwood, T. K., Pigford, R. L., and Wilke, C. R. (1975). *Mass Transfer*. McGraw-Hill, New York.

Todd, D. K. (1980). *Groundwater Hydrology*. John Wiley & Sons, New York.

U.S. EPA. (1989). *Transport and Fate of Contaminants in the Subsurface*. EPA/625/4-89/019. U.S. Environmental Protection Agency Center for Environmental Research Information, Cincinnati, OH.

U.S. EPA. (1991a). *Site Characterization for Subsurface Remediation*. EPA/625/4-91/026. U.S. Environmental Protection Agency Office of Research and Development, Washington, DC.

U.S. EPA. (1991b). *Description of Sampling of Contaminated Soils: A Field Pocket Guide*. EPA 625/12-91/002. U.S. Environmental Protection Agency Center for Environmental Information, Cincinnati, OH.

U.S. EPA. (1991c). *Compendium of ERT Groundwater Sampling Procedures*. EPA/540/P-91/007. U.S. Environmental Protection Agency Office of Emergency and Remedial Response, Washington, DC.

U.S. EPA. (1991d). *Handbook of Suggested Practices for the Design and Installation of Ground-Water Monitoring Wells*. EPA160014-891034. U.S. Environmental Protection Agency Office of Research and Development, Las Vegas, NV.

U.S. EPA. (1992). *RCRA Ground-Water Monitoring: Draft Technical Guidance*. U.S. Environmental Protection Agency Office of Solid Waste, Washington, DC.

U.S. EPA. (1993a). *Use of Airborne, Surface, and Borehole Geophysical Techniques at Contaminated Sites - A Reference Guide.* EPA/625/R-92/007. U.S. Environmental Protection Agency Office of Science Planning and Regulatory Evaluation, Cincinnati, OH.

U.S. EPA. (1993b). *Subsurface Characterization and Monitoring Techniques -- A Desk Reference Guide (Volume I: Solids and Ground Water; Appendices A and B).* EPA/625/R-93/003a. U.S. Environmental Protection Agency Center for Environmental Information, Washington, DC.

U.S. EPA. (1993c). *Subsurface Characterization and Monitoring Techniques: A Desk Reference Guide (Volume II: The Vadose Zone, Field Screening and Analytical Methods; Appendices C and D).* EPA/625/R-93/003b. U.S. Environmental Protection Agency Center for Environmental Research Institute, Washington, DC.

U.S. EPA. (1996). *Soil Screening Guidance: User's Guide. 9355.4–23.* U.S. Environmental Protection Agency Office of Solid Waste and Emergency Response, Washington, DC.

U.S. EPA. (2002a). *Guidance on Choosing a Sampling Design for Environmental Data Collection for Use in Developing a Quality Assurance Project Plan.* EPA/240/R-02-005. U.S. Environmental Protection Agency Office of Environmental Information, Washington, DC.

U.S. EPA. (2002b). *RCRA Waste Sampling Draft Technical Guidance.* EPA-530-D-02-002. U.S. Environmental Protection Agency Office of Solid Waste, Washington, DC.

U.S. EPA. (2002c). *Ground-Water Sampling Guidelines for Superfund and RCRA Project Managers.* EPA 542-S-02-001. U.S. Environmental Protection Agency Office of Solid Waste and Emergency Response, Washington, DC.

U.S. EPA. (2005a). *In-Situ DUOXTM Chemical Oxidation Technology to Treat Chlorinated Organics at the Roosevelt Mills Site, Vernon, CT. Site Characterization and Treatability Report.* U.S. Environmental Protection Agency Superfund Innovative Technology Evaluation, Washington, DC.

U.S. EPA. (2005b). *USEPA Region 9 Technical Guidelines for Accurately Determining Volatile Organic Compound (VOC) Concentrations in Soil and Solid Matrices.* U.S. Environmental Protection Agency Region 9 Quality Assurance Office, San Francisco, CA.

U.S. EPA. (2011). *Environmental Cleanup Best Management Practices: Effective Use of the Project Life Cycle Conceptual Model.* EPA 542-F-11-011. U.S. Environmental Protection Agency Office of Solid Waste and Emergency Response, Washington, DC.

U.S. EPA. (2014). *Soil Sampling. SESDPROC-300-R3.* U.S. Environmental Protection Agency Region 4 Science and Ecosystem Support Division, Atlanta, GA.

U.S. EPA. (2020). *Soil Gas Sampling. LSASDPROC-307-R4.* U.S. Environmental Protection Agency Region 4 Laboratory Services and Applied Science Division, Atlanta, GA.

U.S.G.S. (2017). *Acadian Pontchartrain (ACAD) Groundwater Studies.* https://archive.usgs.gov/archive/sites/la.water.usgs.gov/nawqa/liaison/gwgeneral.htm (Jul. 24, 2020).

Ward, A. D., Trimble, S. W., Burckhard, S. R., and Lyon, J. G. (2014). *Environmental Hydrology.* CRC Press, Boca Raton, FL.

5 Vadose Zone Soil Remediation

5.1 INTRODUCTION

The previous chapters covered the essence of contaminated sites, what regulations affect them, how to evaluate risk, and how to conduct a remedial investigation. If the site investigation concludes that chemicals of concern (COCs) are present in the subsurface in concentrations above acceptable levels, then the remediation of the impacted soil is required.

Many techniques have been developed and used to remediate, remove, and stabilize the impacted soil. These techniques can be applied *in situ* or *ex situ*. *In situ* is a Latin term meaning "in its original place," and *ex situ* is just the opposite, "outside of its original place." These terms are used to describe an action that remediates the soil in its original place and an action that remediates the soil by removing it from its original place and treating it.

Throughout this chapter, when the word "soil" is used, it means the soil in the vadose zone, also known as the unsaturated zone, which is above the groundwater table (see Figure 2.5 in Chapter 2). We focus on the vadose zone in this chapter because this is where contamination starts. When an underground storage tank (UST) leaks or a tanker truck rolls over, the COC first spills into the surrounding soil above the water table. A leak or spill is much easier to remediate in the vadose zone before the COCs migrate into the groundwater. Another way to think of this scenario is that contaminated soil in the vadose zone is the source of groundwater contamination. Therefore, by prioritizing cleanup of the soil, we are eliminating a source of contamination to groundwater.

The remedial objective is to reduce the COC concentrations to below the acceptable cleanup levels. Therefore, this chapter will cover commonly used methods to meet the remedial objective. These are excavation, soil vapor extraction (SVE), bioremediation, chemical oxidation, soil washing, phytoremediation, capping, and solidification and stabilization.

5.2 EXCAVATION

Possibly the simplest and quickest technique to reduce COCs in the vadose zone is the excavation of the soil, followed by disposal or remediation *ex situ*. This technique is simple and quick if the conditions are right; that is, if the contamination is shallow and small enough to be inexpensive and if the excavation does not interfere with existing infrastructure in use, such as buildings, roads, and utilities. Because of the relatively quick time to excavate, excavation of contaminated soil is often conducted during the remedial investigation (RI) phase. Because the RI phase often involves the removal of leaking USTs, excavation can be done during that time.

5.2.1 FUNDAMENTAL CONCEPTS

If contaminated soil is excavated, it needs to be treated or disposed of in a hazardous waste landfill. The excavated soil is typically first stored on-site in stockpiles (Figure 5.1). The amount of excavated soil can be determined by measuring the volumes of the stockpiles. If feasible, the contractor should separate the apparently impacted soil from the clean soil (excavating some clean soil is unavoidable) by putting it into separate piles to save money on the subsequent treatment and disposal costs. Using a portable instrument to detect the COCs in real time, such as a photo-ionization detector (PID, shown in Figure 5.2), flame-ionization detector (FID), or organic vapor analyzer (OVA), helps in making this decision.

The key information needed in excavation projects is listed below.

- Dimensions of the excavated pit (from field measurements)
  - A simple excavated pit would be rectangular in shape.
  - Most excavations are not perfectly rectangular, so the most approximate geometric shape should be considered.
  - Excavations deeper than 5 ft (1.5 m) are considered a caving hazard, meaning that a vertical excavated wall can be unstable enough to cave in on a worker (OSHA 2011). For this reason, excavations deeper than 5 ft (1.5 m) require either sloped walls or shoring of vertical walls.

FIGURE 5.1   An excavated soil stockpile in the town of Pines, Indiana, covered to prevent chemicals from volatilizing into the atmosphere and becoming an inhalation hazard. (Source: U.S. EPA 2020.)

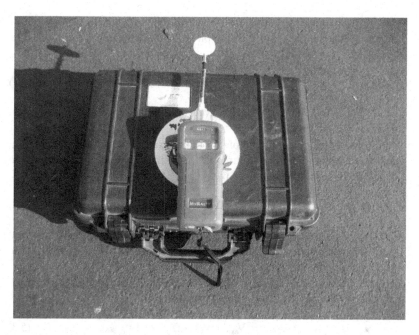

FIGURE 5.2   A portable photo-ionization detector (PID), which measures organic vapors on site. (Photo credit: Michael Shiang.)

- Number and volume of any USTs removed (from drawings or field observation; also refer to Figure 2.1 in Chapter 2 for a photo of a UST removal)
  - A typical UST volume at a gasoline station is approximately 10,000 gallons (38,000 L). However, USTs can come in many sizes and range from 550 to 30,000 gallons (145 to 114,000 L).
- Total bulk density of soil, ρ_t (from a measurement or estimate)
  - The total bulk density of soil is the combined density of the soil grains, water content, and air (see Sections 2.4.2 and 2.7.4).
- Soil fluffy factor (from a field estimate)
  - This factor accounts for the loosening of soil after being excavated from the subsurface. The *in situ* soil is compacted, and the *ex situ* soil is not.
  - A fluffy factor of, say, 1.1, means that the volume of soil increases by a factor of 1.1, or 10%, from *in situ* to the stockpile.
  - The total bulk density of soil in the stockpiles would be smaller than that of *in situ* soil as a result of becoming loose after excavation and having more air between the soil grains.
- The concentration of COCs in soil, *in situ* or in stockpiles.

5.2.2   EXCAVATION CALCULATIONS

To determine how much soil to excavate, how much soil to ship to a disposal site, and what mass of COCs is present in that soil, several calculation steps are useful, as shown below.

Step 1: To determine the volume of the excavation, measure the dimensions of the pit. Calculate the volume of the pit from the measured dimensions. For a rectangular pit,

$$V_{pit} = L \times W \times D \tag{5.1}$$

where V_{pit} is the volume of the excavation, L is the length, W is the width, and D is the depth.

Step 2: To determine the volume of soil in the excavation if USTs are present, determine the number and volume of USTs, then subtract the total volume of the USTs from the volume of the pit.

$$V_{soil} = V_{pit} - V_{USTs} \tag{5.2}$$

Where V_{soil} is the volume of soil in the excavation and V_{USTs} is the volume of USTs, which can be the volume capacity of the UST as an approximation, or the actual volume that the UST occupies in the soil.

Step 3: To determine the volume of soil that has been expanded after excavation, multiply the value from Step 2 by a soil fluffy factor.

$$V_{exp} = V_{soil} \times f_f \tag{5.3}$$

Where V_{exp} is the expanded volume of soil in the stockpile and f_f is the fluffy factor.

Step 4: To determine the mass of soil, multiply the volume of the soil by the total bulk density. There are two ways of doing this: (1) multiply the *in situ* volume by the *in situ* total bulk density ($\rho_{t,in\ situ}$) or (2) multiply the stockpile volume by the *ex situ* total bulk density ($\rho_{t,ex\ situ}$), which is the *in situ* bulk density divided by the fluffy factor.

$$M_{soil} = V_{soil} \times \rho_{t,in\ situ} \tag{5.4}$$

or

$$M_{soil} = \frac{V_{exp} \times \rho_{t,in\ situ}}{f_f} = V_{exp} \times \rho_{t,ex\ situ} \tag{5.5}$$

where M_{soil} is the mass of soil.

Step 5: To determine the mass of COCs present in the soil, multiply an average COC concentration from samples by the volume and total bulk density of soil, as seen in Chapter 2, Eq. 2.4.

$$M_{COC} = X_{COC} \times V_{soil} \times \rho_t \tag{2.4}$$

Where M_{COC} is the mass of COCs in the soil and X_{COC} is the average concentration of COCs in the samples.

The following example demonstrates some of these steps.

Example 5.1: Mass and Volume of Soil Excavated from a Tank Pit

Two 20,000 L (5,000-gallon) USTs and one 23,000 L (6,000-gallon) UST were removed in an excavation that resulted in a tank pit of 15 m×8 m×5.5 m (50′×26′×18′). The excavated soil was stockpiled on site. The total bulk density of soil *in situ* (before excavation) is 1.8 g/cm³ (112 lb/ft³), and that of soil in the stockpiles is 1.5 g/cm³ (94 lb/ft³). Estimate the volume and mass of the excavated soil.

This problem will first be solved in SI units, then in the U.S. system (USCS) of units.

SOLUTION IN SI UNITS:

Step 1. Calculate the volume of the excavation using Eq. 5.1.

$$V_{pit} = 15 \text{ m} \times 8 \text{ m} \times 5.5 \text{ m} = 660 \text{ m}^3$$

Step 2. Calculate the volume of soil by taking the result above and subtracting the volume of the tanks.

$$V_{USTs} = 2 \times 20,000 \text{L} + 23,000 \text{ L} = 63,000 \text{ L}$$

Then, using Eq. 5.2 and a conversion factor:

$$V_{soil} = 660 \text{m}^3 - 63,000 \text{L} \times \frac{\text{m}^3}{1000 \text{L}} = 597 \text{m}^3$$

Step 3. Estimate the volume of the excavated soil. In this example, we were not given a fluffy factor, but we were given both the *in situ* bulk density and the *ex situ* bulk density. We can use these two densities to calculate a fluffy factor:

$$f_f = \frac{\rho_{t,in \, situ}}{\rho_{t,ex \, situ}} = \frac{1.8 \text{ g} / \text{cm}^3}{1.5 \text{ g} / \text{cm}^3} = 1.2$$

Then use Eq. 5.3:

$$V_{exp} = 597 \text{ m}^3 \times 1.2 = 716.4 \text{ m}^3$$

Step 4. Determine the mass of the soil by using the *ex situ* bulk density and multiplying by conversion factors:

$$M_{soil} = V_{exp} \times \rho_{t,ex \, situ} = 716.4 \text{m}^3 \times 1.5 \frac{\text{g}}{\text{cm}^3} \times \left(\frac{100 \text{cm}}{1 \text{m}}\right)^3 \frac{\text{kg}}{1000 \text{g}} = 1,074,600 \text{ kg} \cong 1,100 \text{ tonnes}$$

The volume of excavated soil is approximately 720 m³, and the mass is 1,100 tonnes or metric tons.

SOLUTION IN USCS UNITS:

Step 1. Calculate the volume of the excavation using Eq. 5.1.

$$V_{pit} = 50 \text{ ft} \times 26 \text{ ft} \times 18 \text{ ft} = 23,400 \text{ ft}^3$$

Step 2. Calculate the volume of soil by taking the result above and subtracting the volume of the tanks.

$$V_{USTs} = 2 \times 5,000 \text{ gal} + 6,000 \text{ gal} = 16,000 \text{ gal}$$

Then, using Eq. 5.2 and a conversion factor:

$$V_{soil} = 23,400 \text{ ft}^3 - 16,000 \text{ gal} \times \frac{1 \text{ft}^3}{7.48 \text{ gal}} = 21,261 \text{ ft}^3$$

Step 3. Estimate the volume of the excavated soil. In this example, we were not given a fluffy factor, but we were given both the *in situ* bulk density and the *ex situ* bulk density. We can use these two densities to calculate a fluffy factor:

$$f_f = \frac{\rho_{t,\text{in situ}}}{\rho_{t,\text{ex situ}}} = \frac{112 \text{lb}/\text{ft}^3}{94 \text{lb}/\text{ft}^3} = 1.2$$

Then use Eq. 5.3:

$$V_{exp} = 21,261 \text{ ft}^3 \times 1.2 = 25,513 \text{ ft}^3$$

Step 4. We'll determine the mass of the soil by using the *ex situ* bulk density:

$$M_{soil} = V_{exp} \times \rho_{t,\text{ex situ}} = 25,513 \text{ ft} \times 94 \frac{\text{lb}}{\text{ft}^3} = 2,398,236 \text{ lb}$$

Converting from lb to (short) tons:

$$2,398,236 \text{ lbm} \times \frac{1 \text{ ton}}{2000 \text{ lbm}} = 1199 \text{ (short) tons}$$

The volume of excavated soil is approximately 25,500 ft³, and the mass is 1,200 (short) tons.

DISCUSSION:

1. These final answers have been rounded to 2 or 3 significant figures considering the relative uncertainty inherent in the measurements in the given information.
2. The calculated mass of the excavated soil should be the same whether you use the volume of soil in the tank pit or that in the stockpile.

3. For the SI system of units, one tonne or metric ton = 1,000 kg (equivalent to 2,200 lb), and for the U.S. customary system (USCS) of units, one (short) ton = 2000 lb.

So far, we have performed calculations to determine the volume and mass of soil excavated. But we can use these calculations with additional field and laboratory measurements to estimate the mass of COCs present in that soil. Laboratory methods to determine soil concentrations rely on a person to collect several samples of soil in small jars to represent the entire stockpile. Estimating the mass of COCs in the excavated soil will tell us how much contamination was removed from the subsurface. Figure 5.3 shows how samples can be collected from cores into the piles.

Let's say that three samples were collected from a stockpile, and the concentrations of total petroleum hydrocarbons (TPH) in each one are 1,500 mg/kg, 2,000 mg/kg, and < 100 mg/kg. If we assume that those three samples represent the entire stockpile, then we can use the average of the three concentrations to obtain an overall concentration of TPH. But note that the TPH concentration for one of the three samples is < 100 mg/kg. This means the TPH in that sample is below the laboratory detection limit of 100 mg/kg. There may be no TPH at all or up to 99.9 mg/kg in the sample. There are four approaches commonly taken to deal with value below the detection limit: (1) use the detection limit as the value, (2) use half of the detection limit, (3) use zero, and (4) select a value based on a statistical approach (especially when multiple samples are taken, and a few of them are below the detection limit). The most conservative approach is to use the detection limit as the concentration. Often in tables of technical articles, values below the detection limit are shown as "ND" for non-detect or not detected. It is preferable to show the corresponding detection limits of these samples, such as "ND (<100 mg/kg)."

The next two examples demonstrate the scenario above.

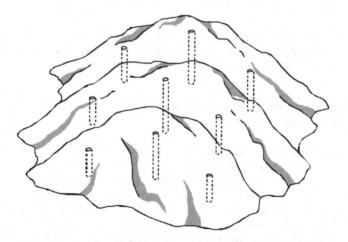

FIGURE 5.3   Cores are used to collect soil samples in stockpiles. (Source: U.S. EPA 2002.)

Example 5.2: COC Concentrations in Excavated Soil

A leaking 1,000-gallon underground storage tank was removed from the subsurface. The excavation resulted in a tank pit of 12' × 12' × 15' (L×W×D), and the excavated soil was stockpiled on site. Five samples were collected from the pile and analyzed for TPH using EPA Method 8015. The bulk densities of the soil are 112 lb/ft³ in place and 93 lb/ft³ in the pile. Assume the density of TPH is 50 lb/ft³.

Based on the laboratory results, an engineer estimated that there were approximately 50 gallons of gasoline distributed throughout the stockpile. What would be the average concentration of TPH in the soil stockpile?

SOLUTION:

The average concentration of TPH is the mass of TPH divided by the mass of soil:

$$\bar{X}_{TPH} = \frac{M_{TPH}}{M_{soil}}$$

The mass of soil is the volume of the excavation (minus the volume of the tank) multiplied by the bulk density in place:

$$M_{soil} = V_{soil} \times \rho_t = \left(12 \text{ ft} \times 12 \text{ ft} \times 15 \text{ft} - 10,000 \text{ gal} \times \frac{\text{ft}^3}{7.48 \text{ gal}} \right) \times 112 \frac{\text{lb}}{\text{ft}^3} = 226,948 \text{ lb}$$

The engineer has estimated that 50 gal of TPH spilled in the soil, so we can use the density of TPH to find its mass, as shown in the numerator of the initial equation:

$$\bar{X}_{TPH} = \frac{M_{TPH}}{M_{soil}} = \frac{50 \text{ gal} \times 50 \frac{\text{lb}}{\text{ft}^3} \times \frac{\text{ft}^3}{7.48 \text{ gal}}}{226,948 \text{ lb}} = 0.00147 \frac{\text{lb}}{\text{lb}} = 0.00147 \frac{\text{kg}}{\text{kg}} = 1,470 \frac{\text{mg}}{\text{kg}}$$

The average TPH concentration is 1,470 mg/kg. Knowing this concentration will eventually help the engineer to determine how to treat or dispose of the soil. Note that we did not need to use the bulk density of the stockpile.

Example 5.3: Estimate the Mass of COCs from an Average Concentration

An engineer has to determine the average concentration of TPH in a stockpile in order to determine how to treat or dispose of the soil. Four soil samples have been collected, and the TPH values in the report are ND (<150), 1400, 2100, and 3700 mg/kg, respectively. The volume of soil in the stockpile is estimated at 2440 ft³, with a stockpile bulk density of 93 lb/ft³. (Yes, mixing units is common because chemical analytical labs typically use SI units and geotechnical labs use USCS units.) What is the average TPH concentration, and what is the estimated mass of TPH in the stockpile, if the density of TPH is 50 lb/ft³?

SOLUTION:

In order to design a solution for the worst-case scenario, use the conservative approach and let the ND concentration be the highest possible, 150 mg/kg. We expect that the average TPH concentration is

$$\bar{X}_{TPH} = \frac{150 + 1,400 + 2,100 + 3,700}{4} = 1,838 \cong 1,800 \text{ mg / kg}$$

The mass of TPH is shown in the equation below, which includes a conversion factor of lb to kg.

$$M_{TPH} = \bar{X}_{TPH} \times V_{soil} \times \rho_t = 1838 \frac{mg}{kg} \times 2440 \text{ ft}^3 \times 93 \frac{lb}{ft^3} \times \frac{0.4536 \text{ kg}}{lb} = 1.89 \times 10^8 \text{ mg}$$

$$= 190 \text{ kg of TPH}$$

The average concentration is 1,800 mg/kg, and the mass is 190 kg of TPH. Note that these final figures are rounded to two significant figures to match the number of significant figures in the given information. Also note the four significant figures for the TPH concentration as an intermediate step to find the final mass of TPH.

5.3 SOIL VAPOR EXTRACTION

Soil vapor extraction (SVE), also known as soil venting, *in situ* vacuum extraction, *in situ* volatilization, or soil vapor stripping, is a popular remediation technique for soil impacted by volatile organic compounds (VOCs). The process strips VOCs in the vapor phase from the impacted soil by inducing an air flow through the impacted zone. The air flow is created by a vacuum pump (often called a "blower") through one of more vapor extraction wells drilled into the vadose zone.

5.3.1 FUNDAMENTAL CONCEPTS

As a vacuum pump suctions the vapor away from the voids within the soil grains of the vadose zone, fresh air is naturally (through passive venting wells or air infiltration) or mechanically (through air-injection wells) introduced to refill the void. This flux of fresh air will: (1) disrupt the existing partition of the COCs among the voids, soil moisture, and soil grain surfaces by promoting volatilization of the dissolved COCs and desorption of the adsorbed COCs; (2) provide oxygen to indigenous microorganisms for biodegradation of the COCs; and (3) carry away the toxic metabolic by-products generated from the biodegradation process (Figure 5.4). The extracted air is usually laden with VOCs and brought to the ground surface by the vacuum pump. Treatment of the extracted vapor is normally required before release into the ambient air. The design for the VOC-laden air treatment is covered in Chapter 7.

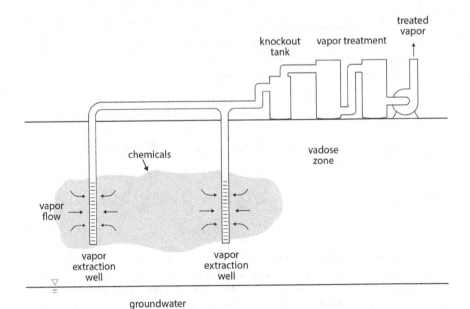

FIGURE 5.4 A schematic of an SVE system showing, from left to right, two vapor extraction wells, a knockout tank, and a treatment unit, and, to the right, the vacuum pump that extracts the vapor and suctions it through the treatment unit. The vapor exits the pump in a vertical pipe and enters the atmosphere.

Several concepts used in SVE have been covered in Chapter 2, for example, vapor pressure (P^{vap}), partial pressure (P_A), Henry's law constant (H), the octanol-water partition coefficient (K_{ow}), organic content (f_{oc}), porosity (ϕ), and bulk density of the soil (ρ_t).

5.3.2 DESIGN OF SOIL VAPOR EXTRACTION SYSTEMS

Major components of a typical SVE system (Figures 5.4 and 5.5) include vapor extraction wells (Figure 5.6), vacuum pumps, a moisture-removal device (the knock-out tank), vapor collection piping and ancillary equipment, and the vapor treatment system.

The most important parameters for the preliminary design of an SVE system are the extracted VOC concentration, air flow rate, radius of influence of the venting well, number of wells required, the location of the wells, and the size of the vacuum pump.

Expected Vapor Concentrations

VOCs in the vadose zone may be present in four phases: (1) in the soil moisture due to dissolution, (2) on the surface of the soil grain due to adsorption, (3) in the voids among the soil grains due to volatilization, and (4) as the free product.

FIGURE 5.5   A portable SVE unit consisting of a vacuum pump (left), knockout tank (front), granular activated carbon (GAC) treatment unit (rear), control panel (gray), with vacuum gauges, dilution valves, flow meters, and sample ports. (Photo credit: Michael Shiang.)

If the COCs are present in the free product phase, the vapor concentration in the voids can be estimated from Raoult's law:

$$P_A = \left(P^{vap}\right)\left(x_A\right) \tag{2.10}$$

where P_A is the partial pressure of compound A in the vapor phase, P^{vap} is the vapor pressure of compound A as a pure liquid, and x_A is the mole fraction of compound A in the liquid phase.

Raoult's law is previously described in Section 2.5.3. The partial pressure calculated from Eq. 2.10 represents the upper limit of the COC concentration in the extracted vapor from an SVE system. The actual concentration is lower than this upper limit because (1) not all the extracted air passes through the impacted zone, and (2) mass transfer is not 100% efficient. Nevertheless, this concentration serves as a starting point for estimating the initial vapor concentration at the beginning of an SVE project. Initially, the extracted vapor concentration will be relatively constant if LNAPL is present above the groundwater table. As soil venting continues, the free product phase will disappear. The extracted vapor concentration will then begin to drop and become dependent on the partitioning of the COCs among the remaining three phases. As the air flows through the pores and transports the COCs, the COCs

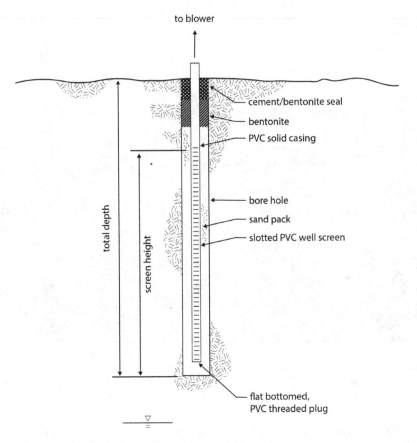

FIGURE 5.6   A typical vertical SVE well construction detail. (Modified from U.S. EPA 2017.)

dissolved in the soil moisture have a stronger tendency to volatilize from the liquid into the pores. Simultaneously, some COCs will desorb from the soil grain surface and enter into the soil moisture. Consequently, the concentrations in all three phases decrease as the venting process progresses, thereby, remediating the vadose zone.

These phenomena describe common observations at sites that contain a single type of COC. SVE has also been widely used for sites impacted by a mixture of compounds, such as gasoline. For these cases, the vapor concentration decreases continuously from the start of venting, and a period of constant vapor concentration in the beginning phase of the project may not exist. This can be explained by the fact that each compound in the mixture has a different vapor pressure. Thus, the more volatile compounds tend to leave the free product, the moisture, and the soil surface earlier than the less volatile ones. Table 2.2 shows the molecular weights of VOCs and their vapor pressures.

To estimate the initial concentration of the extracted vapor in equilibrium with the free-product phase, the following procedure can be used:

Step 1. Obtain the vapor pressure, P^{vap}, of the COC (e.g., from Table 2.2 or the NIOSH Pocket Guide (NIOSH 2007)).

Step 2. Determine the mole fraction, x_A, of the COC in the free product. For a pure compound, set $x_A = 1.0$, or 100%. For a mixture, follow the procedure in Section 2.5.3.

Step 3. Use Eq. 2.10 to estimate the partial pressure, P_A. This is also the vapor concentration, in units of pressure. The discussion in Example 2.7 shows the equivalent values of units of concentration and pressure.

Step 4. Convert the concentration by volume into a mass concentration, if needed, using Eqs. 2.5 and 2.6.

Example 5.4: Estimate the Maximum Gasoline Vapor Concentration

Two sites are impacted by accidental gasoline spills. The spill at the first site happened recently, while the spill at the other site occurred three years ago, so gasoline is considered to be weathered, meaning that the chemical properties have changed. Estimate the maximum gasoline vapor concentration at each site. The vapor pressures at 20°C are:

$$P^{vap}{}_{\text{fresh gasoline}} = 0.34 \text{ atm}$$

$$P^{vap}{}_{\text{weathered gasoline}} = 0.049 \text{ atm}$$

The molecular weights are:

$$MW_{\text{fresh gasoline}} = 95 \text{ g/mol}$$

$$MW_{\text{weathered gasoline}} = 111 \text{ g/mol}$$

SOLUTION:

Determine the mole fraction of each COC. Because both sites are affected with free product and no chemical mixtures, then

$$x_{\text{fresh gasoline}} = 1.0$$

$$x_{\text{weathered gasoline}} = 1.0$$

Next, use Eq. 2.10 to estimate the vapor concentration. Because the mole fraction of each COC is 1.0, the vapor pressure is equal to the vapor concentration or partial pressure.

$$P_{\text{fresh gasoline}} = P^{vap}{}_{\text{fresh gasoline}} \times x_{\text{fresh gasoline}} = 0.34 \text{ atm} \times 1.0 = 0.34 \text{ atm}$$

$$P_{\text{weathered gasoline}} = P^{vap}{}_{\text{weathered gasoline}} \times x_{\text{weathered gasoline}} = 0.049 \text{ atm} \times 1.0 = 0.049 \text{ atm}$$

Next, convert the partial pressure concentration to a vapor concentration (see the discussion in Example 2.7). First, convert on a volume basis:

$$G_{\text{fresh gasoline}} = P_{\text{fresh gasoline}} \times 10^6 \text{ ppmV / atm} = 0.34\,\text{atm} \times 10^6 \text{ ppmV / atm} = 340{,}000 \text{ ppmV}$$

$$G_{\text{weathered gasoline}} = P_{\text{weathered gasoline}} \times 10^6 \text{ppmV / atm} = 0.049 \text{ atm} \times 10^6 \text{ ppmV / atm}$$
$$= 49{,}000 \text{ ppmV}$$

Then, convert the concentration of each type of gasoline from a volume basis to a mass basis using Eq. 2.5, at 1 atm and 20°C.

$$G_{\text{fresh gasoline}} \left[\text{mg / m}^3 \right] = G_{\text{fresh gasoline}} \left[\text{ppmV} \right] \times MW_{\text{fresh gasoline}} \Big/ \left(V_{\text{1 mole ideal gas}} \left[\text{m}^3 \right] \times 1{,}000 \right)$$

$$G_{\text{fresh gasoline}} \left[\text{mg/m}^3 \right] = 340{,}000 \times 95 \text{ g/mol/}(0.02405 \text{ m}^3 \times 1{,}000) = 1.34 \times 10^6 \text{ mg/m}^3$$

And

$$G_{\text{weathered gasoline}} \left[\text{mg/m}^3 \right] = G_{\text{weathered gasoline}} \left[\text{ppmV} \right] \times MW_{\text{weathered gasoline}} \Big/ \left(V_{\text{1 mole ideal gas}} \left[\text{m}^3 \right] \times 1{,}000 \right)$$

$$G_{\text{weathered gasoline}} \left[\text{mg/m}^3 \right] = 49{,}000 \times 111 \text{ g/mol/} (0.02405 \text{ m}^3 \times 1{,}000) = 2.26 \times 10^5 \text{ mg/m}^3$$

The estimated maximum vapor concentrations are 1.3×10^6 mg/m^3 (1.3 kg/m^3) for fresh gasoline and 2.3×10^5 mg/m^3 (0.23 kg/m^3) for weathered gasoline. Note how the lower vapor pressure of weathered gasoline results in a lower vapor concentration compared to fresh gasoline. Therefore, SVE is more desirable for fresh gasoline spills because the chemicals will more easily volatilize.

Radius of Influence and Pressure Profile

Selecting the number and locations of vapor extraction wells is one of the major tasks in the design of *in situ* SVE systems. The decisions are typically based on the radius of influence (R_I).

The radius of influence can be defined as the distance from the extraction well to where the pressure drawdown is very small. The pressure drawdown is the difference between atmospheric pressure and the pressure at a distance r from the extraction well, see Eq. 5.6 and Figure 5.7.

$$\text{Pressure drawdown at distance r} = P_{\text{atm}} - P_r \tag{5.6}$$

The most accurate and site-specific R_I values are measured on-site. This section will describe how to determine the R_I during a pilot test. A pilot test for SVE is a small-scale test with just one vapor extraction well connected to a vacuum pump (blower) and several vapor monitoring wells. Pressure gauges are installed at the extraction and monitoring wells (Figure 5.8). The pressure drawdown data at the extraction well and the monitoring wells can be plotted as a function of the radial distance from the extraction well on a semi-log plot to determine the R_I of the extraction well. The R_I is commonly chosen to be the distance where the pressure drawdown is less than 1% of the vacuum in the extraction well.

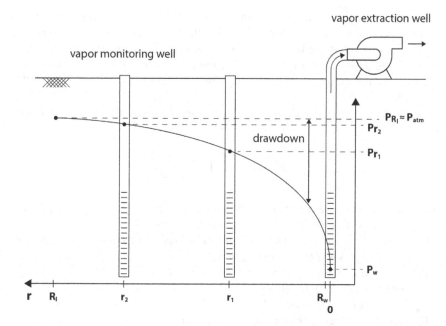

FIGURE 5.7   A profile view of a vapor extraction well (to the right) and two vapor monitoring wells. The vertical axis represents pressure, and the horizontal axis represents the radial distance from the extraction well. The curve represents the pressure in the soil at a continuous distance from the extraction well. The pressure drawdown arrows show the difference between atmospheric pressure and the pressure at any distance from the extraction well.

FIGURE 5.8   A pressure gauge connected to a vapor monitoring well. (Photo credit: Michael Shiang.)

The field test data can also be analyzed by using the flow equation, which describes the subsurface air flow. The subsurface is usually heterogeneous, and the air flow through it can be very complex. As a simplified approximation, a flow equation was derived for a fully confined radial gas flow system in a permeable formation having uniform and constant properties (Johnson et al. 1990a & 1990b; Kuo et al. 1991).

For the steady-state radial flow, the pressure distribution in the subsurface can be derived as

$$P_r^2 - P_w^2 = \left(P_{RI}^2 - P_w^2\right)\frac{\ln\left(r/R_w\right)}{\ln\left(R_I/R_w\right)}\tag{5.7}$$

where the boundary conditions are $P = P_w$ at $r = R_w$ and $P = P_{atm}$ at $r = R_I$, given the following variable definitions and their illustration in Figure 5.7.

P_r = absolute pressure (not gauge pressure) at a distance r from the extraction well
P_w = absolute pressure at the extraction well
P_{RI} = absolute pressure at the radius of influence
r = distance from the extraction well
R_w = radius of the extraction well
R_I = radius of influence from the extraction well

It is useful to remember the definition and relationship between absolute pressure and gauge pressure. Absolute pressure is the real pressure in a system. Gauge pressure is what is measured by a gauge instrument, for example, a tire gauge, and is the difference between absolute pressure and atmospheric pressure:

$$P_{gauge} = P_{abs} - P_{atm}\tag{5.8}$$

When P_{abs} is less than P_{atm}, then a vacuum exists. In that case, P_{gauge} is negative, leading to the term "negative pressure". It is important to use the correct pressure, P_{gauge} or P_{abs} in the given equations.

The pressure applied at the extraction well is a vacuum, that is, an absolute pressure lower than atmospheric pressure. Sometimes one may have difficulty with this concept, and it is useful to know other examples of vacuum in life, such as a vacuum cleaner (where a motor creates a vacuum to suction dust from the floor) and drinking from a straw (where the lungs create a vacuum to suction the drink into the mouth). Figure 5.7 shows these pressure and distance relationships using a vapor extraction well and two vapor monitoring wells where the pressure can be measured. Any two of these locations can be used in Eq. 5.7. Note that the pressure at the extraction well is at its lowest, meaning a higher vacuum, because of the proximity of the pump. As the distance from the applied vacuum increases, the pressure gets closer to atmospheric pressure.

Eq. 5.7 can be used to determine the R_I of a vapor extraction well, if the pressure drawdown data of the extraction well and that of a monitoring well (or drawdown data of two monitoring wells) are known. As shown in Eq. 5.7, the flow rate and the permeability of the formation are not necessary for this approximation. The R_I can also be estimated from the vapor extraction rate and the pressure drawdown data in the extraction well, as will be shown in the subsection Vapor Flow Rates.

If no pilot tests are conducted, an estimate could be made based on previous experience. R_I values ranging from 9 ft (3 m) to 125 ft (38 m) are reported in the literature (Crawford et al. 2012) and typical pressures in the extraction wells range from 0.90 to 0.95 atm. Shallower wells, a less permeable subsurface, and a lower applied vacuum in the extraction well generally correspond to smaller R_I values.

Example 5.5: Determine the Radius of Influence Using Pressure Drawdown Data

Determine the radius of influence of a vapor extraction well using the following information:

- Gauge pressure at the extraction well = 48 inches of water vacuum = - 48 inches of water pressure
- Gauge pressure at a monitoring well 30 ft away from the extraction well = 8 inches of water vacuum = - 8 inches of water pressure
- Diameter of the extraction well = 4 inches

SOLUTION:

Because the pressure data is given in gauge pressure, first those numbers must be converted to absolute pressure using Eq. 5.8. One atmosphere is equal to 406.6 inches of water (Recall from Chapter 2, Discussion in Example 2.7, that one atmosphere is equal to 29.9 inches of mercury. Mercury has a specific gravity of 13.6, meaning that it is 13.6 times denser than water. So a column of water of 406.6 inches is equal to 406.6/13.6 = 29.9 inches of mercury.)

$$P_{abs} = P_{gauge} + P_{atm}$$

Absolute pressure at the extraction well = P_w = $406.6 + (-48) = 358.6$ inches H_2O

Absolute pressure at the monitoring well = P_r = $406.6 + (-8) = 398.6$ inches H_2O

We can define R_I as the location where the drawdown is equal to 1% of the vacuum in the extraction well. This is a more conservative estimate than the definition above. In this case:

Drawdown at the extraction well = $P_{atm} - P_w = 406.6 - 358.6 = 48$ inches H_2O

and

P_{RI} = Absolute pressure at R_I = $406.6 - (48)(1\%) = 406.12$ inches H_2O

Then Eq. 5.7 becomes

$$398.6^2 - 356.6^2 = \left(406.12^2 - 356.6^2\right)\frac{\ln\left[30/(2/12)\right]}{\ln\left[R_I/(2/12)\right]}$$

Solving for the radius of influence:

$$R_I = 85 \text{ ft}$$

For comparison, if R_I is the location where P is equal to the atmospheric pressure, then R_I can be found by using Eq. 5.7 and appropriate unit conversions:

$$398.6^2 - 358.6^2 = \left(406.6^2 - 358.6^2\right)\frac{\ln\left[30/(2/12)\right]}{\ln\left[R_I/(2/12)\right]}$$

$$R_I = 91 \text{ ft}$$

When the R_I is defined as the location where the vacuum is 1% of that at the extraction well, the radius of influence is 85 ft, which is a more conservative answer than the 91 ft when we let the R_I occur when atmospheric pressure is achieved.

Vapor Flow Rates

In designing an SVE system, an engineer needs an estimate of the vapor flow rate that will be generated given the site conditions. For this, it is necessary to first calculate the radial Darcy velocity, which is the vapor velocity flowing through the soil pores and entering a well. This velocity, u_r, is the vapor flow velocity at a radial distance r away from the extraction well and has been determined for homogeneous soil systems by (Johnson et al. 1990a):

$$u_r = \left(\frac{K}{2\mu}\right)\frac{\left[\dfrac{P_w}{r\ln\left(\dfrac{R_w}{R_I}\right)}\right]\left[1-\left(\dfrac{P_{RI}}{P_w}\right)^2\right]}{\left\{1+\left[1-\left(\dfrac{P_{RI}}{P_w}\right)^2\right]\dfrac{\ln\left(\dfrac{r}{R_w}\right)}{\ln\left(\dfrac{R_w}{R_I}\right)}\right\}^{0.5}} \tag{5.9}$$

where

K = permeability of the formation, units of darcy (1 darcy = 10^{-8} cm^2 = 10^{-12} m^2)

μ = viscosity of air. Typically, 0.018 centipoise is used (1 poise = 0.1 Pa×s = 100 centipoise = 0.1 N×s/m^2)

and the other variables are as in Eq. 5.7.

The velocity at the wellbore, u_w, can be found by replacing r with R_w in the above equation:

$$u_w = \left(\frac{K}{2\mu}\right)\left[\frac{P_w}{R_w \ln\left(\frac{R_w}{R_I}\right)}\right]\left[1 - \left(\frac{P_{RI}}{P_w}\right)^2\right]$$ (5.10)

The volumetric vapor flow rate entering the extraction well, Q_w, can be found as the product of the surface area of the cylinder of the screened interval and the wellbore velocity:

$$Q_w = (2\pi R_w H)(u_w)$$ (5.11)

where H is the height of the screened interval of the extraction well. Substituting u_w and simplifying:

$$Q_w = H\left(\frac{\pi K}{\mu}\right)\left[\frac{P_w}{\ln(R_w/R_I)}\right]\left[1 - \left(\frac{P_{RI}}{P_w}\right)^2\right]$$ (5.12)

To convert the vapor flow rate entering the well to the flow rate discharged to the atmosphere (Q_{atm}), where $P = P_{atm} = 1$ atm, Eq. 5.13 can be used:

$$Q_{atm} = \left(\frac{P_{well}}{P_{atm}}\right)Q_w$$ (5.13)

Example 5.6: Estimate the Extracted Vapor Flow Rate of a Vapor Extraction Well

A vapor extraction well (4-inch diameter) was installed at a site. The pressure in the extraction well is 0.9 atm and the radius of influence of this soil venting well has been determined to be 50 ft.
    Calculate (a) the steady-state flow rate entering the well per unit height of the well screen, (b) the vapor flow rate in the well, and (c) the vapor rate at the extraction pump discharge by using the following additional information:

- Permeability of the formation = 1 darcy
- Well screen height = 20 ft
- Viscosity of air = 0.018 centipoise
- Temperature of the formation = 20°C

SOLUTION:

First, we will convert units to the same system of units. We will choose to convert to SI here because some of the constants are given in SI.

Diameter of the well: $d = 4\text{ inches} \times \dfrac{1\text{ foot}}{12\text{ inches}} \times \dfrac{0.3048\text{ m}}{1\text{ ft}} = 0.1016\text{ m}$. Therefore,

the radius of the well is 0.0508 m

Pressure at the well: $P_w = 0.9\text{ atm} \times \dfrac{101,325\text{ N}/\text{m}^2}{1\text{ atm}} = 91,193\text{ N}/\text{m}^2$

Radius of influence: $R_I = 50\text{ ft} \times \dfrac{0.3048\text{ m}}{\text{ft}} = 15.24\text{ m}$

Permeability: 1 darcy $= 10^{-12}$ m² (from the variable description of Eq. 5.9)

Well screen height: $20\text{ ft} \times \dfrac{0.3048\text{ m}}{\text{ft}} = 6.096\text{ m}$

Viscosity: $0.018\text{ centipoise} \times \dfrac{0.1\dfrac{\text{N}\cdot\text{s}}{\text{m}^2}}{100\text{ centipoise}} = 1.8 \times 10^{-5}\text{ N}\cdot\text{s}/\text{m}^2$ (from the

variable description of Eq. 5.9)

(a) The velocity at the wellbore, u_w, can be found using Eq. 5.10. We can define R_I as the location where the drawdown is equal to 1% of the vacuum in the extraction well.

Drawdown at the monitoring well $= P_{atm} - P_w = 101,325\dfrac{\text{N}}{\text{m}^2} - 91,193\dfrac{\text{N}}{\text{m}^2}$

$$= 10,132\text{ N}/\text{m}^2$$

and

$P_{RI} =$ Absolute pressure at $R_I = 101,325 - (10,132)(1\%) = 101,222\text{ N}/\text{m}^2$

$$u_w = \left(\dfrac{10^{-12}\text{ m}^2}{2 \times 1.8 \times 10^{-5}\text{ N}\cdot\text{s}/\text{m}^2}\right)\left[\dfrac{91,193\text{ N}/\text{m}^2}{0.0508\,m\ln\left(\dfrac{0.0508\text{ m}}{15.24\text{ m}}\right)}\right]\left[1 - \left(\dfrac{101,222\text{ N}/\text{m}^2}{91,193\text{ N}/\text{m}^2}\right)^2\right]$$

$$= 0.00203\text{ m}/\text{s}$$

The wellbore velocity is 0.00203 m/s, which may be better pictured if expressed in units of m/h: 7.3 m/h.

(b) The vapor flow rate entering the well can be found using Eq. 5.11.

$$Q_w = 2\pi(0.0508\text{ m})(6.096\text{ m})\left(0.00203\dfrac{\text{m}}{\text{s}}\right) = 0.0039\dfrac{\text{m}^3}{\text{s}} = 0.24\dfrac{\text{m}^3}{\text{min}} = 14\text{ m}^3/h$$

(c) The vapor flow rate at the extraction pump discharged at atmospheric pressure can be found using Eq. 5.13:

$$Q_{atm} = \left(\dfrac{0.9}{1}\right)14\dfrac{\text{m}^3}{\text{h}} = 12.3 = 13\text{ m}^3/h$$

Note that the pressure ratio in this equation was expressed in units of atm. If expressed in different units, the ratio would still be the same because the units in the numerator and denominator cancel out.

Discussion: Using consistent units in Eq. 5.12 is very important. In this example, the pressure is in N/m^2, the distance in m, the permeability in m^2, and the viscosity in $N \times s/m^2$. Therefore, the velocity is in m/s.

Number of Vapor Extraction Wells

The sections above describe characteristics of a single vapor extraction well. But most of the time, multiple wells are necessary to cover the entire contaminated vadose zone. The number of extraction wells must cover the entire plume. In other words, the entire impacted area should be within the influence of the wells (see Figure 5.9). The number of wells, N_{wells}, is

$$N_{wells} = \frac{1.2 \left(A_{plume} \right)}{\pi R_I^2} \tag{5.14}$$

where A_{plume} is the area of the plume, and R_I is the radius of influence of an individual vapor extraction well. The factor 1.2 is arbitrarily chosen to account for the overlapping of the influence areas among the wells.

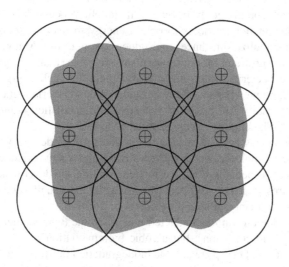

$\oplus$ **vapor extraction well**

FIGURE 5.9 A demonstration of how vapor extraction wells are arranged so that their areas of influence overlap and cover the entire contaminated area (in gray).

5.3.3 Variations of Soil Vapor Extraction

The same principles of SVE can be applied to variations of this technology. Example variations are listed below.

- SVE with horizontal wells. Improved drilling technology can be used to install diagonal or horizontal SVE wells in locations where the vadose zone contamination would be hard to reach with vertical wells, for example, under buildings.
- Dual-phase extraction or multi-phase extraction. This variation of SVE extracts not only soil vapor but groundwater (in the case of dual-phase) and groundwater and floating free product (in the case of multi-phase). In these cases, inside the vapor extraction well, a smaller diameter tube, or straw, is installed into the top of the groundwater table. The vacuum pump that typically would be suctioning vapor, would then also slurp the water and free product at the top of the aquifer, thereby limiting the movement of contaminated water and free product.

5.4 SOIL BIOREMEDIATION

5.4.1 Fundamental Concepts

Bioremediation, in general, involves microorganisms degrading organic COCs. Living organisms obtain energy through a metabolic process. Just like people eat as part of their metabolic process to obtain energy to live, microorganisms must consume and metabolize substances to produce energy to live. Many organic COCs can be "food," or substrate, for microorganisms. In general, microorganisms in bioremediation are bacteria, which are microscopic single-celled organisms. These organisms undergo metabolism and respiration. When bacteria metabolize organic COCs, they generate waste products. Effective bioremediation requires that those waste products be less toxic than the original COCs.

Bioremediation has been successfully applied to degrade petroleum hydrocarbons, solvents, pesticides, wood preservatives, and other organic compounds. Bioremediation is especially effective for remediating low-level residual contamination. It requires relatively inexpensive materials and usually does not generate residual wastes requiring additional treatment or disposal.

Soil bioremediation can be conducted under aerobic (with oxygen) or anaerobic (absence of oxygen) conditions, but aerobic bioremediation is more popular. The final waste products of complete aerobic biodegradation of hydrocarbons are carbon dioxide, water, and microbial cell mass. In anaerobic bioremediation, organic COCs may be metabolized into methane.

Microorganisms require moisture, oxygen (or absence of oxygen for anaerobic biodegradation), nutrients, and a suitable set of environmental factors to grow. The environmental factors include appropriate pH and temperature, and the absence of toxic conditions. Table 5.1 summarizes the critical conditions needed for microbes to perform bioremediation.

TABLE 5.1
Critical Conditions for Bioremediation (U.S. EPA 1991)

Environmental Factor	Optimum Conditions
Available soil water	25–85% water holding capacity
Oxygen	Aerobic metabolism: > 0.2 mg/L dissolved oxygen
	air-filled pore space > 10% by volume
	Anaerobic metabolism: oxygen concentration < 1% by volume
Redox potential	Aerobes and facultative anaerobes: > 50 millivolts
	Anaerobes: < 50 millivolts
Nutrients	Sufficient N, P, and other nutrients
	(suggested C:N:P molar ratio of 120:10:1, where moles of C are from
	the organic chemical and N and P are added nutrients)
pH	5.5–8.5 (for most bacteria)
Temperature	15–45°C (for mesophiles)

Bioremediation can be carried out following two primary processes: biostimulation and bioaugmentation. Bioremediation is considered a destructive technology in that an organic molecule is degraded, or broken down, into smaller compounds. The goal is for a harmful COC molecule to be degraded into less harmful ones. Because the process is done via living microorganisms, it is called biodegradation. We will focus on biostimulation since it is the most often used bioremediation process. But, briefly, bioaugmentation involves the use of microbial cultures that have been specially bred for degradation of the COCs and survival under severe environmental conditions.

Biostimulation is the process of stimulating indigenous (in nature) microorganisms to metabolize COCs. The concept behind biostimulation is that in soil, a large amount and variety of microorganisms exist. Among those microorganisms, some are especially suited to degrade certain organic COCs. If, out of all existing types of microorganisms in the soil, those that can degrade a particular type of COC can be stimulated, then this is the practice of biostimulation.

The stimulants for biostimulation are often substances that promote more efficient respiration in those microorganisms. For example, the most common stimulant for aerobic bioremediation is oxygen. Many organic COCs are biodegraded by aerobic microorganisms. If oxygen is added to the soil, then those aerobic microorganisms will undergo efficient respiration and consume the COCs as part of their metabolic process. Common biostimulants are oxygen and nutrients, primarily ammonium and phosphate, as listed in Table 5.1.

Bioremediation can happen *in situ* and *ex situ*. *In situ* treatment enhances the natural microbial activity of undisturbed soil in place to decompose organic COCs. The most common application of *in situ* vadose zone bioremediation is bioventing, discussed in Section 5.4.2.

Ex situ soil bioremediation is typically performed using one of three systems: (1) static soil pile, (2) in-vessel (inside a tank), and (3) slurry bioreactor. The static soil pile is the most popular *ex situ* format. This approach treats soil stockpiled on the site with perforated pipes embedded in the piles as the conduit for biostimulants to be applied. In aerobic bioremediation, often a forced air supply is introduced into the pipes to increase the levels of oxygen in the pile. To minimize fugitive air emissions and potential secondary contamination from leachate, the stockpiles are usually covered on the top and lined at the bottom and an air vacuum is applied to induce the air/oxygen into the piles. A similar process can be carried out in a vessel. In a slurry bioreactor, impacted soil is mixed in a tank with a nutrient solution under controlled operating conditions (i.e., optimal pH, temperature, dissolved oxygen, and mixing).

Cometabolism occurs when microorganisms growing on one compound produce an enzyme that chemically transforms another compound on which they cannot grow. In particular, microorganisms that degrade methane (methanotrophic bacteria) have been found to produce enzymes that can initiate the oxidation of a variety of chlorinated organic compounds under anaerobic conditions.

Many books on microbiology and bioremediation can complement the theory of bioremediation in more depth. The focus of this chapter is the science and engineering principles for applying bioremediation in the field.

5.4.2 Applications

Calculations for Oxygen Biostimulation

For aerobic bioremediation, the oxygen involved in the biological activity is often supplied through the oxygen in the air. Oxygen is approximately 21% by volume in the atmosphere (approx. 279,300 mg/m^3 at 20°C and 1 atm). On the other hand, oxygen is sparingly soluble in water. At 20°C, the saturated dissolved oxygen (DO_{sat}) concentration in water is about 9 mg/L, or 9,000 mg/m^3.

Let us use the following simplified scheme to demonstrate the required oxygen to convert C into CO_2:

	C	+	O_2	→	CO_2
Moles:	1		1		1
Mass (g/mol or lb/lb-mol):	12		32		44

The above equation illustrates that each mole of carbon requires one mole of the oxygen molecule, or every 12 grams of carbon requires 32 grams of oxygen, a ratio of 2.67. Other elements in the COCs, such as hydrogen, nitrogen, and sulfur, would also demand oxygen for bioremediation. For example, the theoretical amount of oxygen required to aerobically biodegrade benzene, C_6H_6, can be found as:

	C_6H_6	+	$7.5O_2$	→	$6CO_2$	+	$3H_2O$
Moles:	1		7.5		6		3
Mass (g/mol or lb/lb-mol):	78		240		264		54

This indicates that each mole of benzene requires 7.5 moles of the oxygen molecule, or every 78 grams of carbon requires 240 grams of oxygen, a ratio of 3.08, which is larger than the 2.67 based on pure carbon. Using benzene as the basis, it means that every gram of hydrocarbon requires about 3 grams of oxygen for aerobic degradation. It should be noted that this is the theoretical ratio based on the stoichiometric relationship. A larger amount of oxygen would be needed to account for the inefficiency of this reaction in the natural environment. Using the ratio of 1 g COC to 3 g oxygen, the amount of oxygen in an aqueous solution can only support the biodegradation of COCs at a concentration of 3 mg/L or less if the water is saturated with 9 mg/L of DO at 20°C. The next example shows how oxygen supplied by air is more efficient than oxygen supplied by water for bioremediation.

Example 5.7: Compare the Oxygen Concentration in Air to that of Water

To see how the oxygen concentration in air compares to the saturated oxygen concentration in water (9 mg/L), determine the mass concentration of oxygen in ambient air at 20°C when the volume concentration is approximately 21%. Express the answer in units of mg/L, g/L, and lb/ft³.

SOLUTION:

A 21% volumetric concentration of oxygen is equal to 210,000 ppmV because 210,000/1,000,000 = 0.21 or 21%. Eq. 2.5 can be used to convert ppmV to a mass concentration:

$$G_{oxygen}\left(in\ \frac{mg}{m^3}\right) = 210,000\ ppmV\left(\frac{32}{24.05}\right) \quad at\ 20°C\ and\ 1\ atm$$

$$G_{oxygen} = 279,300\ mg/m^3$$

Now convert 279,300 mg/m³ to the required units:

$$G_{oxygen} = 279,300\ \frac{mg}{m^3} \times \frac{1m^3}{1000\ L} = 279\ mg/L$$

$$G_{oxygen} = 279\ \frac{mg}{L} \times \frac{g}{1000\ mg} = 0.279\ g/L$$

$$G_{oxygen} = 0.279\ \frac{g}{L} \times \frac{lb}{453.59\ g} \times \frac{28.32\ L}{ft^3} = 0.0174\ lb/ft^3$$

The oxygen concentration in ambient air, 279 mg/L, is much higher than the saturated dissolved oxygen (DO) concentration in water, approximately 9 mg/L at 20°C. For *in situ* bioremediation, it's important to remember properties of soil, where COCs are located and where air flows to promote microbial activity. Sections 2.7.2 through 2.7.4 and 2.8 describe these properties. The next example combines soil properties with bioremediation.

Example 5.8: Determine the Mass of Oxygen for Complete Biodegradation

After a gasoline spill, the vadose zone soil of a fueling station averages a concentration of 3,500 mg/kg of benzene. The air in the subsurface is relatively stagnant. The total bulk density (ρ_t) of the soil is 1.8 g/cm³, the degree of moisture content in soil (S_w) is 30%, and the porosity (ϕ) is 40%.

On the basis of 1 m³ of soil, determine: (a) the mass of benzene present; (b) the theoretical mass of oxygen required for complete biodegradation; (c) the mass of oxygen in the soil moisture; (d) the mass of oxygen in the air voids; and (e) the additional mass of oxygen needed for complete biodegradation of benzene.

SOLUTION:

(a) The mass of benzene in 1.0 m³ of soil can be found using Eq. 2.4:

$$M_{TPH} = XV\rho_t = 3500\,\frac{mg}{kg} \times 1\,m^3 \times 1.8\,\frac{g}{cm^3} \times \left(\frac{100\,cm}{m}\right)^3 \times \left(\frac{1\,kg}{1000\,g}\right)$$

$$= 6.3 \times 10^6\ mg = 6.3\ kg$$

(b) The theoretical mass of oxygen required for complete degradation can be assumed based on the 3.08 ratio of oxygen to carbon in benzene.

$$Oxygen\ requirement = (3.08)(6.3\ kg) = 19.4\ kg\ of\ O_2$$

(c) For a review of soil moisture, refer to Section 2.7.3. The mass of oxygen in the soil moisture (assuming that the moisture is saturated with oxygen and the saturated dissolved oxygen concentration in water at 20°C is approximately 9 mg/L):

$$Volume\ of\ soil\ moisture = V_{sm} = V\phi S_w = \left(1\,m^3\right)(0.40)(0.30) = 0.12\,m^3$$

$$Mass\ of\ oxygen\ in\ soil\ moisture = V_{sm}(DO) = \left(0.12\,m^3\right)\left(9\,\frac{mg}{L}\right)\left(\frac{1000\,L}{1\,m^3}\right)$$

$$= 1,080\ mg = 1.08\ g$$

(d) Mass of oxygen in the air voids (assuming that the oxygen concentration is the same as in ambient air, 21% by volume or 279 mg/L from Example 5.7):

$$Volume\ of\ the\ air\ void, V_{air\,void} = V\phi(1-S_w) = \left(1\,m^3\right)(0.40)(1-0.30) = 0.28\,m^3$$

$$Mass\ of\ oxygen\ in\ air\ void = V_{air\,void}(G_{oxygen}) = \left(0.28\,m^3\right)\left(279\,\frac{mg}{L}\right)\left(\frac{1000\,L}{1\,m^3}\right)$$

$$= 78,120\ mg = 78.1\,g$$

(e) To find the additional mass of oxygen needed for biodegradation, first add the oxygen available in the soil moisture and in the air voids:

$$\text{Mass of oxygen available} = 1.08\ g + 78.1\ g = 79.2\ g$$

Then, compare with the oxygen requirement:

$$79.2\ g \ll 19.4\ kg$$

$$\text{Additional mass of oxygen needed} = 19,400\ g - 79.2\ g = 19,300\ g$$

Discussion: This example shows that the oxygen available just in the soil moisture and in the air voids is not enough for bioremediation and that additional oxygen would need to be added to biodegrade benzene.

Bioventing

Bioventing is a bioremediation process that is similar to soil vapor extraction (SVE), wherein vapor extraction wells are built and connected to a vacuum pump (Figure 5.10). However, the purpose of bioventing is not to extract the vapor from the vadose zone; it is to mobilize air into the contaminated area and promote *in situ* bioremediation in the form of biostimulation. Oxygen already present in the air stream within the soil flows towards the contaminated area and stimulates the biodegradation of the COCs. Air flow rates in bioventing are much lower than in SVE.

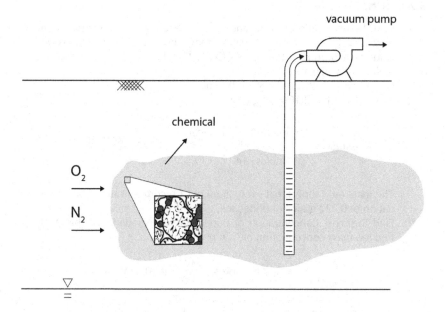

FIGURE 5.10 A bioventing system with a vapor extraction well and vacuum pump promoting air movement that biodegrades the COCs.

The preferred COCs are heavy hydrocarbons with low vapor pressures, which will not volatilize significantly, but biodegrade *in situ*.

The following example demonstrates how bioventing is used for biodegradation while extracting some vapors. This example also demonstrates an activity that monitors the effectiveness of a bioventing system in progress.

Example 5.9: Estimate the Amount of Biodegradation of a COC from Bioventing

Bioventing is used to remediate a site impacted by diesel fuel. Recent extracted air samples have average concentrations of total petroleum hydrocarbons (TPH, used to represent diesel) and CO_2 of 500 ppmV and 50,000 ppmV, respectively. CO_2 at 50,000 ppmV is equivalent to 5%, which is much greater than the 400 ppmV (0.04%) of CO_2 in the atmosphere. This indicates that biodegradation of diesel is occurring, and CO_2 is resulting from the biodegradation. Because there is no specific formula for TPH, use dodecane ($C_{12}H_{26}$) to represent TPH, and therefore diesel. The following equation shows the stoichiometric amounts involved in the biodegradation of dodecane with oxygen to produce carbon dioxide and water:

$$C_{12}H_{26} + 18.5O_2 \rightarrow 12CO_2 + 13H_2O$$

Estimate the mass percentage of dodecane removal by biodegradation from this bioventing process.

SOLUTION:

(a) Find the mass ratio of dodecane to CO_2 necessary for biodegradation. According to the chemical equation above, every mole of $C_{12}H_{26}$ biodegraded will produce 12 moles of CO_2. To find the mass ratio, first find the molecular weights.
MW $[C_{12}H_{26}] = 12(12) + 1(26) = 170$ g/mol
MW $[CO_2] = 12 + 2(16) = 44$ g/mol

$$\text{Mass ratio} = \frac{1\,\text{mole} \times 170\,\frac{g}{mol}\,\text{dodecane}}{12\,\text{moles} \times 44\,\frac{g}{mol}\,CO_2} = 0.322\,\frac{g\,\text{dodecane}}{g\,CO_2}$$

The mass ratio shows that every gram of CO_2 comes from the biodegradation of 0.322 grams of dodecane.

(b) Find the mass concentration of dodecane biodegraded. For this, first, find the mass concentration of CO_2 using Eq. 2.5.

$$G_{CO_2}\left(in\,\frac{mg}{m^3}\right) = G_{CO_2}\,(in\,ppmV)\frac{MW}{24.05}\ \text{at }20°C\text{ and }1\,atm$$

$$G_{CO_2}\left(in\,\frac{mg}{m^3}\right) = (50,000\,ppmV)\frac{44}{24.05} = 91,480\,\frac{mg}{m^3}\,CO_2$$

Based on the ratio of 0.322 found in part (a), the concentration of dodecane biodegraded is:

$$\left(91,480 \text{ mg CO}_2\right)\left(0.322 \frac{\text{mg dodecane}}{\text{mg CO}_2}\right) = 29,460 \text{ mg dodecane}$$

Therefore, 29,460 mg/m³ of dodecane are biodegraded, resulting in 91,480 mg/m³ of CO_2.

(c) To answer the main question of finding the percentage of dodecane removed via biodegradation, take the mass of dodecane in the air sample and compare with how much was biodegraded. For the mass of dodecane, follow the same procedure as in part (b).

$$G_{dodecane}\left(\text{in } \frac{\text{mg}}{\text{m}^3}\right) = \left(500 \text{ ppmV}\right)\frac{170}{24.05} = 3,534 \frac{\text{mg}}{\text{m}^3} \text{ dodecane}$$

The concentration of 3,534 mg/m³ of dodecane is the amount that remains in the air sample and was not biodegraded. Knowing from part (b) that 29,460 mg/m³ biodegraded, then the percentage of dodecane removed via biodegradation is

$$\%\text{biodegraded} = \frac{29,460 \text{ mg}/\text{m}^3}{29,460 \frac{\text{mg}}{\text{m}^3} + 3,534 \frac{\text{mg}}{\text{m}^3}} = 0.89 = 89\%$$

Therefore, 89% of the dodecane has theoretically biodegraded. This also means that the remaining 11% was also removed, as evidenced by the air sample, via volatilization.

Discussion: This calculation is made using chemical dodecane as a surrogate TPH, which in turn is a surrogate for diesel. This provides an approximate answer. If the formula for the particular diesel spilled is available and can be used, then the answer will be more accurate.

Nutrient-Enhanced Biostimulation

When contaminated soil is excavated, bioremediation can be conducted *ex situ*, in stockpiles or tanks. *Ex situ* bioventing can be one of the technologies, but *ex situ* soil is also able to be bioremediated using nutrients. Nutrients necessary for microbial activity usually naturally exist in soils. However, with the elevated level of organic COCs in contaminated soil, additional nutrients are often needed to support the activity of indigenous microbes to stimulate the bioremediation. The nutrients to enhance microbial growth are assessed primarily on the nitrogen and phosphorus requirements. As shown in Table 5.1, a typical suggested C:N:P molar ratio is 120:10:1. This means that every 120 moles of carbon biodegradation by microbes requires 10 moles of nitrogen and 1 mole of phosphorus to stimulate microbes. A lab-based feasibility study is usually conducted to determine the exact C:N:P ratio for a particular site. Nutrients are often dissolved in water first and are then applied to the soil by spraying or irrigation.

To determine the nutrient requirements, the following procedure can be used:

Step 1: Determine the mass of the organics present in the impacted soil.

Step 2: Find the moles of the COC by dividing the mass of COC by its molecular weight.

Step 3: Multiply the moles of COC from step 2 by the number of C in the compound's formula.

Step 4: Determine the moles of nitrogen and phosphorus needed using the optimal C:N:P ratio. For example, if the ratio is C:N:P = 120:10:1, then

$$\text{Moles of nitrogen needed} = (\text{moles of carbon present}) \times (\text{N:C ratio})$$

$$\text{Moles of phosphorus needed} = (\text{moles of carbon present}) \times (\text{P:C ratio})$$

Step 5: Determine the mass of nutrient needed.
Information needed for this calculation:
- Mass of the organic COCs
- Chemical formula of the COCs
- Optimal C:N:P ratio
- Chemical formula of the nutrients

Example 5.10: Estimate the Mass and Cost of Nutrients Needed for Bioremediation

The results of a feasibility study indicate that the excavated soil in a stockpile is suitable for on-site aboveground *ex situ* bioremediation. The study also determined that the optimum C:N:P molar ratio is 100:10:1. Two types of nutrients are being considered: ammonium sulfate $((NH_4)_2SO_4)$ as the N source (priced at $3/lb) and tri-sodium phosphate $(Na_3PO_4 \cdot 12H_2O)$ as the P source (priced at $9/lb). Estimate the mass and cost of each nutrient (in lb) needed to remediate the impacted soil. Other parameters are given below:

- Initial mass of gasoline in the piles = 158 kg
- Formula for gasoline (assumed) = C_7H_{16}

SOLUTION:

(a) We are first being asked for the mass of each nutrient, given a molar ratio. So first we need to find the moles of C, N, and P to then convert to the mass of the nutrient compounds that have N and P.
We are given the mass of gasoline, which we can convert to moles:
MW gasoline = 7×12 g/mol + 16×1 g/mol = 100 g/mol
Moles of gasoline = Mass/MW = 158 kg/(100 g/mol) = 1.58 kmol = 1.58×10^3 mol
Now we have to find the moles of C in gasoline. Since there are 7 C atoms in a gasoline molecule, then

Moles of $C = 1.58$ kmol $\times 7 = 11.06$ kmol

Now that we have the moles of C, we use the ratio 100:10:1 to find the required moles of N and P.

Required moles of $N = 11.06$ kmol $\times (10/100) = 1.106$ kmol

Required moles of $P = 11.06$ kmol $\times (1/100) = 0.1106$ kmol

(b) Now that we have the moles of N and P, we need to find the moles of each associated compound.

There are 2 moles of N in 1 mole of $(NH_4)_2SO_4$, and there is 1 mole of P in 1 mole of $Na_3PO_4 \cdot 12H_2O$.

Required moles of $(NH_4)_2SO_4 = 1.106/2 = 0.553$ kmol of ammonium sulfate

Required moles of $Na_3PO_4 \cdot 12H_2O = 0.1106$ kmol

(c) Now that we have the moles of compounds to be added, we find the mass:

Mass $(NH_4)_2SO_4 = MW \times$ moles $= [(14 + 1 \times 4) \times 2 + 32 + 16 \times 4]$kg/kmol $\times$ 0.553 kmol $= 73$ kg $= 161$ lb

Mass $Na_3PO_4 \cdot 12H_2O = MW \times$ moles $= [(23)(3) + 31 + 16(4) + 12(18)] \times$ 0.1106 kmol $= 42$ kg $= 92.5$ lb

(d) These compounds are powders purchased typically in bags of 50 lb each, so we would need 200 lb of $(NH_4)_2SO_4$ and 100 lb of $Na_3PO_4 \cdot 12H_2O$, which would cost a total of $200 \times 3 + 100 \times 9 = \$1,500$. This would be the cost of the nutrient only and does not include the cost of tanks, water, pumps, and labor for injecting these nutrients into the excavated soil.

5.5   *IN SITU* CHEMICAL OXIDATION

In situ chemical oxidation (ISCO) in the vadose zone typically involves the introduction of a chemical oxidant into an open excavation to degrade COCs into less harmful compounds, usually water and carbon dioxide. Chemical oxidants can also be introduced into an excavated stockpile. ISCO is also a groundwater remediation technology (see Chapter 6). Transforming COCs into less harmful chemicals in an excavation prevents the COCs from infiltrating deeper into the vadose zone and the groundwater table where they can travel with groundwater. Consequently, ISCO applied to the vadose zone can shorten the time it takes to clean up a site (U.S. EPA 2006). ISCO, also applied to groundwater, follows the same principles described in this section.

5.5.1   FUNDAMENTAL CONCEPTS

Advantages of ISCO over more conventional technologies, such as SVE, include less waste generation and shorter time of remediation. These advantages result in savings in the cost of materials, waste disposal, and labor and energy for long-term operation and maintenance (ITRC 2005).

In practice, ISCO systems are composed of equipment to store and inject the oxidants. The oxidant can be a liquid or a gas. Figure 5.11 is an example showing a white cylindrical chemical storage tank, tubing, and a pump to spread the oxidant solution into an excavation.

FIGURE 5.11   An ISCO system installed at a dry cleaning facility. (Source: U.S. EPA 2012.)

In the broadest use of term, ISCO refers to the use of chemical oxidation for the in-place remediation of contaminants. In practice, the term is often used in describing the process of injection or direct mixing of reactive chemical oxidants with the soil and groundwater for the purpose of destroying (oxidizing) chemical contaminants in place. Within the context of soil mixing, this term refers to the use of chemical oxidants delivered and mixed with the soil, via soil mixing, for the purpose of in-place treatment. The most common type of soil mixing used for ISCO is large-diameter single auger soil mixing, but the technique can also be accomplished using excavator buckets and rotary tools. Soil mixing offers numerous technical advantages over alternative oxidant delivery methods, with the largest advantage being that close contact between the oxidant and contaminants is guaranteed with soil mixing, independent of soil lithology. Other methods like direct injection are not effective in certain soil types and, even in soil types that are conducive to direct injection, the injection program can have variable effectiveness in achieving oxidant-contaminant contact.

The following subsections describe the oxidants and the oxidation process used in ISCO.

Common Oxidants

Various oxidants that can be injected into the subsurface for ISCO. The most common ones are:

- Permanganate (MnO_4^-), often as potassium or sodium permanganate ($KMnO_4$ or $NaMnO_4$)
- Hydrogen peroxide (H_2O_2)
- Fenton's reagent (hydrogen peroxide + ferrous iron, $H_2O_2 + Fe^{2+}$)
- Ozone (O_3)
- Persulfate ($S_2O_8^{2-}$), often as sodium persulfate ($Na_2S_2O_8$)

The oxidant needs to be persistent, or stable, in the subsurface for it to have enough time to reach and effectively react with and degrade the COCs. Permanganate can persist for months, persulfate for hours to weeks, and hydrogen peroxide, ozone, and Fenton's reagent can only persist for minutes to hours. In the case of persulfate, hydrogen peroxide, and ozone, their free radicals are the chemicals generally considered to be responsible for the transformation of COCs. These free radicals are intermediates with an unpaired electron that react very quickly and persist for very short periods (less than 1 second). So the persistence of the oxidants in the subsurface is important for the release and reactions of the short-lived free radicals. Permanganate-based ISCO is more fully developed than the other forms of oxidants (EPA 2006).

The Oxidation Process

Oxidation is half of a chemical reaction called reduction–oxidation reaction, *redox* reaction in short. Oxidation is a half-reaction where there is a loss of electrons, while reduction is a half-reaction where there is a gain of electrons. When these two half-reactions are added together, they form a complete redox reaction.

Consider the chemical species in a chemical equation to understand the terminology. The species gaining electrons is the oxidizing agent and oxidizes the species that give up the electrons. The reactant that loses electrons is the reducing agent that reduces the reactant that receives the electrons (Ong and Kolz 2007). Oxygen is an oxidant, as implied by the name. Being an oxidant, then it is reduced, gaining electrons and becoming negatively charged. A half-reaction does not have to have oxygen to be an oxidation reaction, nor does a chemical species need to contain oxygen to be an oxidant. But it's good to remember that an oxidant in a redox reaction reacts like oxygen: it gains electrons.

Two common memory tricks to remember these reactions are:

- LEO the lion goes GER! Loss of electrons: oxidation (reactant is oxidized); gain of electrons: reduction (reactant is reduced).
- OIL RIG. Oxidation is a loss of electrons (the oxidized reactant loses electrons); reduction is a gain of electrons (the reduced reactant gains electrons).

In a chemical oxidation process, the COCs are oxidized, and the oxidant is reduced. The reaction involves electron transfers in which the oxidant will serve as a terminal electron acceptor (TEA) by accepting the electrons from the COCs.

The degree of oxidation is measured by a standard oxidation potential (units of volts). Some oxidants are stronger than others, as shown in Table 5.2, and are more effective at ISCO. Note that all the oxidants in this table have a stronger oxidation potential than that of oxygen. Other factors that influence the effectiveness of ISCO are reaction rates, temperature, pH, reactant concentration, catalysts, reaction by-products, and impurities in the system (ITRC 2005).

TABLE 5.2
Oxidant Strengths (Source: Siegrist et al. 2001)

Chemical Species	Standard Oxidation Potential (volts)	Strength Relative to Chlorine (Chlorine = 1)
Hydroxyl radical ($\cdot$OH)[1]	2.8	2.0
Sulfate radical ($\cdot SO_4^-$)	2.5	1.8
Ozone (O_3)	2.1	1.5
Sodium persulfate ($Na_2S_2O_8$)	2.0	1.5
Hydrogen peroxide (H_2O_2)	1.8	1.3
Permanganate (MnO_4^-)	1.7	1.2
Chlorine (Cl_2)	1.4	1.0
Oxygen (O_2)	1.2	0.9

1. This radical can be formed when ozone and hydrogen peroxide decompose.

Below are the reduction half-reactions of the common oxidants. One can identify a chemical reaction as a half-reaction because of the presence of electrons, e^-, as a chemical species. A reduction half-reaction will have electrons on the left side of the yield sign. In the examples below, the oxidants on the left are being reduced, gaining the electrons on the left side of the equation, and therefore oxidizing the COCs. The COCs are not shown here but will be shown later in complementary oxidation half-reactions.

$$MnO_4^- + 4H^+ + 3e^- \rightarrow MnO_2 + 2H_2O \tag{5.15}$$

$$H_2O_2 + 2H^+ + 2e^- \rightarrow 2H_2O$$

$$2 \cdot OH^- + 2H^+ + 2e^- \rightarrow 2H_2O$$

$$O_3 + 2H^+ + 2e^- \rightarrow O_2 + H_2O$$

$$S_2O_8^{2-} + 2e^- \rightarrow 2SO_4^{2-}$$

$$\cdot SO_4^- + e^- \rightarrow SO_4^{2-}$$

$$O_2 + 4e^- \rightarrow 2O^{2-}$$

The above equations show that each mole of hydroxyl radical ($\cdot$OH) or sulfate radical ($\cdot SO_4^-$) can gain one mole of electrons. Each mole of hydrogen peroxide, ozone, of persulfate can gain two moles of electrons. Each mole of permanganate can gain three moles of electrons, and each mole of oxygen can gain four moles of electrons. Table 5.3 tabulates the amount of oxidant needed to transfer one mole of electrons. For a given mass of COC, a smaller oxidant amount would be needed for oxidants that transfer more electrons per unit mass (e.g., oxygen). However, this is not an indicator of whether the reaction can occur.

TABLE 5.3

Amount of Oxidant Needed to Transfer One Mole of Electrons (from Kuo 2014)

	Electrons accepted	Molecular weight	Moles of electrons accepted per unit mass of oxidant
Potassium permanganate	3	158	0.0190
Hydrogen peroxide	2	34	0.0588
Ozone	2	48	0.0417
Sodium persulfate	2	238	0.0084
Oxygen	4	32	0.1250

5.5.2 DETERMINING OXIDANT QUANTITIES

To come up with a reaction equation for the oxidation of a COC, we need to have an oxidation half-reaction. Let's use PCE (C_2Cl_4) as an example COC:

$$C_2Cl_4 + 4H_2O \rightarrow 2CO_2 + 4Cl^- + 8H^+ + 4e^- \qquad (5.16)$$

If permanganate will be used as the oxidant, then we combine Eq. 5.15 with Eq. 5.16 in such a way to cancel out the electrons in the half-reaction. If we add together Eqs. 5.15 and 5.16, then we have to multiply Eq. 5.15 by 4 and Eq. 5.16 by 3, as shown below.

$$4MnO_4^- + 16H^+ + 12e^- \rightarrow 4MnO_2 + 8H_2O$$

$$3C_2Cl_4 + 12H_2O \rightarrow 6CO_2 + 12Cl^- + 24H^+ + 12e^-$$

$$3C_2Cl_4 + 12H_2O + 4MnO_4^- + 16H^+ + 12e^-$$

$$\rightarrow 6CO_2 + 12Cl^- + 24H^+ + 12e^- + 4MnO_2 + 8H_2O$$

Simplifying, we get the final complete redox reaction showing the oxidant (MnO_4^-) and the oxidized COC (C_2Cl_4) on the left side of the yield sign:

$$3C_2Cl_4 + 4MnO_4^- + 4H_2O \rightarrow 6CO_2 + 12Cl^- + 4MnO_2 + 8H^+ \qquad (5.17)$$

Eq. 5.17 shows that the stoichiometric requirement to oxidize PCE is 4/3 mole of permanganate per mole of PCE. Using the same approach, the oxidation of TCE (C_2HCl_3), DCE ($C_2H_2Cl_2$) and vinyl chloride (C_2H_3Cl) can be derived as (EPA 2006):

$$2CHCl_3 + 2MnO_4^- \rightarrow 2CO_2 + 3Cl^- + 2MnO_2 + H^+$$

$$3C_2H_2Cl_2 + 8MnO_4^- \rightarrow 6CO_2 + 6Cl^- + 8MnO_2 + 2OH^- + 2H_2O$$

$$3C_2H_3Cl + 10MnO_4^- \rightarrow 6CO_2 + 3Cl^- + 10MnO_2 + 7OH^- + H_2O$$

As shown, the stoichiometric requirements for TCE, DCE, and vinyl chloride are 1, 8/3, and 10/3 moles of permanganate per mole of COC, respectively. If other oxidants are used, the stoichiometric requirements would change, based on the oxidants listed in Table 5.3. For example, the stoichiometric requirement of sodium persulfate will be 1.5 times that of potassium permanganate because one mole of permanganate can accept 3 moles of electrons, while one mole of persulfate can only accept 2 moles of electrons.

In addition to the oxidant demand from COCs, the added oxidants will also be lost due to subsurface reactions unrelated to the oxidation of COCs, often referred to as the natural oxidant demand (NOD). NOD stems from reactions with organic and inorganic chemical species that are naturally present in the subsurface. Consequently, the total oxidant demand should be the sum of the NOD and the demand from target COCs as:

$$\text{Total Oxidant Demand} = \text{Natural Oxidant Demand} + \text{Demand} \atop \text{from Target COCs} \qquad (5.18)$$

NOD almost always exceeds the oxidant demand from target COCs. NOD has a significant impact on determining if the ISCO is economically feasible and in calculating the applied oxidant dose. Bench and/or pilot-scale testing should be conducted to determine the NOD for a project.

Example 5.11: Estimate the Mass of Oxidant Necessary for Chemical Oxidation

The soil in an excavation is impacted by PCE at a concentration of 5,000 mg/kg. Chemical oxidation will be used for remediation, and two oxidants are being considered. Determine the mass of potassium permanganate and sodium persulfate needed to be delivered to the excavation

SOLUTION:

(a) For potassium permanganate:
    Find the molar concentrations of the species of interest, so start with the MW of each species.

MW of PCE (C_2Cl_4) = $12 \times 2 + 35.5 \times 4 = 166$ g/mol
MW of potassium permanganate ($KMnO_4$) = $39 + 55 + 16 \times 4 = 158$ g/mol
Find the molar concentrations:
Molar concentration of PCE =

$$5,000 \frac{mg}{kg} = 5.0 \frac{g}{kg} = \frac{5.0 g/kg}{166 g/mol} = 3.01 \times 10^{-2} \frac{mol\ PCE}{kg\ soil}$$

As shown in Eq. 5.17, the stoichiometric requirement to oxidize PCE is 4/3 mole permanganate per mole of PCE.

$$\text{Molar concentration of } KMnO_4 = \frac{4\ moles\ KMnO_4}{3\ moles\ PCE} \times 3.01 \times 10^{-2} \frac{mol\ PCE}{kg\ soil}$$

$$= 4.02 \times 10^{-2} \frac{mol\ KMnO_4}{kg\ soil}$$

Now convert this molar concentration to a mass concentration:
Concentration of $KMnO_4$

$$= 4.02 \times 10^{-2} \frac{mol\ KMnO_4}{kg\ soil} \times 158 \frac{g\ KMnO_4}{mol} = 6.35 \frac{g\ KMnO_4}{kg\ soil}$$

(b) For sodium persulfate:
Find the molar concentrations of the species of interest, so start with the MW of each species.
MW of sodium persulfate ($Na_2S_2O_8$) = $23 \times 2 + 32 \times 2 + 16 \times 8 = 238$ g/mol
As shown in Table 5.3, the stoichiometric requirement to oxidize PCE with sodium persulfate is 3/2 times that of potassium permanganate.
Molar concentration of $Na_2S_2O_8$

$$= \frac{3\ moles\ Na_2S_2O_8}{2\ moles\ KMnO_4} \times 4.02 \times 10^{-2} \frac{mol\ KMnO_4}{kg\ soil} = 6.02 \times 10^{-2} \frac{mol\ Na_2S_2O_4}{kg\ soil}$$

Now convert this molar concentration to a mass concentration:
Concentration of $Na_2S_2O_8$ =

$$= 6.02 \times 10^{-2} \frac{mol\ Na_2S_2O_4}{kg\ soil} \times 238 \frac{g\ Na_2S_2O_4}{mol} = 14.3 \frac{g\ Na_2S_2O_8}{kg\ soil}$$

For each kg of soil, 6.35 g of $KMnO_4$ are needed or 14.3 g $Na_2S_2O_8$ are needed. The choice of the oxidant would be based on cost, availability, and performance during pilot studies.

5.6 SOIL WASHING

5.6.1 FUNDAMENTAL CONCEPTS

The majority of the organic and inorganic COCs contained in soil is adsorbed to fine soil grains (i.e., clay or silt) that have large specific surface areas (soil surface area per soil volume) with large amounts of organic matter adsorbed to them. These fine

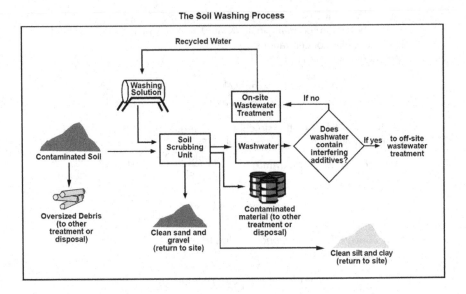

FIGURE 5.12 The soil washing process. (Source: U.S. EPA 1996.)

particles are often mixed with sand and gravel. Sand and gravel are much larger in size, and, because of their mineral structure, they only loosely adsorb COCs. In this section, soil washing is discussed. This technology uses solvents to extract or separate COCs from the soil matrix.

Soil washing is a water-based washing process, in which the COC-impacted soil is excavated, put into a tank for washing, and then removed and returned to the site (Figure 5.12). The major removal mechanisms include desorption of COCs from the soil grains, consequent dissolution into the washing fluid, and/or suspension of the clay and silt particles with bound COCs into the washing fluid. The COCs are readily washed off from sand and gravel, which often account for a large portion of the soil matrix. Separation of the sand and gravel from the heavily impacted clay and silt particles greatly reduces the volume of impacted soil. Soil washing makes further treatment or disposal easier.

Various chemicals can be added to the aqueous solution to enhance the desorption or dissolution of the COCs. For example, an acidic solution is often used to extract heavy metals from the impacted soil. The addition of chelating agents can help the dissolution of heavy metals into the aqueous solution. Chelating agents are chemicals that render toxic metals chemically inert. The addition of surfactants can help the dissolution of organics. Surfactants, also known as surface-active agents, are chemicals that lower the surface tension between two substances. Examples are detergents, foaming agents, and dispersants.

Two other techniques that are similar to soil washing are solvent extraction and soil flushing. Solvent extraction is similar to soil washing, except that solvents, rather than aqueous solutions, are employed to extract organic COCs from the soil. Commonly used solvents are alcohol, liquefied propane and butane, and supercritical

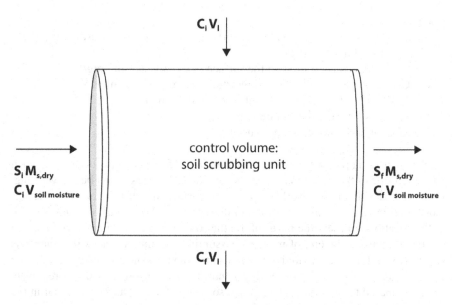

FIGURE 5.13 Mass inputs and outputs in relation to the soil scrubbing unit.

fluids. Soil flushing differs from soil washing and solvent extraction in that it is an *in situ* process in which water or a solvent flushes the impacted zone to desorb or dissolve the COCs. The flushed liquid, or elutriate, is then collected from the wells or drains for further treatment.

5.6.2 Design of a Soil Washing System

A mass balance equation can be written to relate the mass of the COC in the soil before washing and the mass of COC in the spent washing fluid and the mass of the COC in the soil after washing:

Mass of COC in the soil before washing

= Mass of COC in the spent fluid after washing

+ Mass of the COC in the soil after washing

Figure 5.13 shows the mass balance in relation to the soil scrubbing unit, where all mass inputs are shown as arrows into the unit, and all mass outputs are arrows out of the unit. Remember that mass can be expressed as a concentration multiplied by the media volume.

Expressing the mass inputs and outputs from Figure 5.12 as an equation results in Eq. 5.19.

$$S_i M_{s,dry} + C_i V_{soil\ moisture,\ before\ washing} = S_f M_{s,dry} + C_f V_l + C_f V_{soil\ moisture,\ after\ washing} \quad (5.19)$$

where

S_i = initial COC concentration on the surface of the soil before washing (mg/kg)

$M_{s,dry}$ = mass of dry soil (kg)

S_f = final COC concentration on the surface of the washed soil (mg/kg)

C_i = COC concentration in the soil moisture before washing (mg/L)

C_f = COC concentration in the spent washing fluid (mg/L)

$V_{soil\ moisture}$ = volume of the soil moisture (L)

V_l = volume of the washing liquid used (L)

Note that using the units shown, each term in Eq. 5.19 is in units of mg. The term on the left side of the equation represents the total COC mass in soil before washing, which includes the mass adsorbed on the soil surface and that dissolved in the soil moisture (shown as the middle section of the equation). The terms in the last section of the equation represent the mass left on the soil surface and the mass dissolved in the liquid phase at the end of washing. Assuming the mass in the soil moisture is relatively small when compared to mass adsorbed on the soil surface before washing (i.e., $C_i V_{soil\ moisture,\ before\ washing} \ll S_i M_{s,dry}$) and the mass in soil moisture is relatively small compared to the sum of the mass adsorbed on the soil surface and that in the washing fluid after washing (i.e., $C_f V_{soil\ moisture,\ after\ washing} \ll S_f M_{s,dry} + C_f V_{soil\ moisture}$), then Eq. 5.19 can be simplified to

$$S_{initial} M_{s,dry} = S_f M_{s,dry} + C_f V_l \qquad (5.20)$$

These two assumptions are valid when the soil before washing is relatively dry and/ or the COC is relatively hydrophobic.

If an equilibrium condition is achieved at the end of the washing, the COC concentration on the soil and that in the liquid can be related by the partition equation described in Eq. 2.19 and rewritten here with different subscripts:

$$S_f = K_p C_f \qquad (5.21)$$

where K_p is the partition equilibrium constant. Solving Eq. 5.19 for C_f, inserting S_f/K_p for C_f in Eq. 5.21, solving for S_f/S_i and rearranging, we get the relationship between the final and initial concentrations of COC in soil:

$$\frac{S_f}{S_i} = \frac{1}{1 + \dfrac{V_l}{M_{s,dry} K_p}} \qquad (5.22)$$

In addition, if we consider X to be the COC concentration in the soil sample (including COC adsorbed to the soil surface and dissolved in the humidity among the soil grains), we can make the approximation that

$$\frac{S_f}{S_i} \approx \frac{X_f}{X_i}$$

Therefore, Eq. 5.22 can be expressed as

$$\frac{X_f}{X_i} \approx \frac{1}{1 + \dfrac{V_l}{M_{s,dry} K_p}} \tag{5.23}$$

The relationship among mass of soil before washing ($M_{s,wet}$), mass of dry soil ($M_{s,dry}$), dry bulk density (ρ_b), and total bulk density (ρ_t) can be found from the following linear relationship:

$$\frac{M_{s,dry}}{M_{s,wet}} \approx \frac{\rho_b}{\rho_t} \tag{5.24}$$

As described in Section 2.8, the values of S (the adsorbed COC concentration on the soil surface) and X (the COC concentration of the soil sample, including solid surface plus moisture) are relatively similar for compounds like benzene. However, the ratio of the X and S values is essentially the ratio of the dry bulk density and total bulk density for very hydrophobic compounds, such as pyrene.

As we've seen in Eqs. 5.19–5.24, the efficiency of soil washing depends on many parameters covered in Chapter 2: the soil–water partition coefficient K_p, and dry and total bulk density of soil. You may remember from Chapter 2 that other information is needed to find K_p: the fraction of organic carbon in soil/aquifer matrix (f_{oc}), the organic carbon partition coefficient (K_{oc}), and the octanol-water partition coefficient (K_{ow}). This makes sense because how efficiently a COC is washed from soil depends on its affinity to the soil. Therefore, a COC with a high K_{oc} will adhere to soil more than a COC with a low K_{oc} and will, therefore, be more difficult to wash from the soil. The next example will combine those parameters from Chapter 2 with those just introduced in this section.

Example 5.12: Estimate the Final COC Concentrations after Soil Washing

A sandy subsurface contains 500 mg/L of 1,2-dichloroethane (DCA) and 500 mg/L of pyrene. Soil washing is proposed to remediate the soil. A batch washer that can accommodate 1,000 kg of soil is designed. For each batch of operation, 1,000 gallons of clean water are used as the washing fluid. Determine the final concentrations of these two COCs in the washed soil. Use the following parameters, which were introduced in Chapter 2:

- Dry bulk density of soil $= 1.6$ g/cm³
- Total bulk density of soil $= 1.8$ g/cm³
- Fraction of organic carbon of soil, $f_{oc} = 0.005$
- $K_{oc} = 0.63 K_{ow}$

SOLUTION:

We will use Eq. 5.23 solved for X_f, but first, we need to find the parameters to use in the equation.

$$X_f \approx \frac{X_i}{1 + \dfrac{V_l}{M_{s,dry} K_p}}$$

The following parameters are already known:

$X_{i,DCA} = 500$ mg/L; $X_{i,pyrene} = 500$ mg/L; $M_{s,wet} = 1,000$ kg; $\rho_t = 1.8$ g/cm³; $\rho_b = 1.6$ g/cm³; $V_l = 1,000$ gallons $= 3,785$ L (this conversion is done because all other parameters are shown in metric units).

We have to find $M_{s,dry}$ and K_p to solve the equation.

To find $M_{s,dry}$, use Eq. 5.24:

$$M_{s,dry} \approx \frac{\rho_b}{\rho_t} M_{s,wet}$$

$$M_{s,dry} \approx \frac{1.6}{1.8}(1,000 \text{ kg}) = 889 \text{ kg}$$

To find K_p for DCA and pyrene:

From Eq. 2.21, $K_p = f_{oc}K_{oc}$. We know that $f_{oc} = 0.005$. So we need to find K_{oc}. From the given information, $K_{oc} = 0.63K_{ow}$. To find K_{ow}, we look up the values in Table 2.2. So, starting form K_{ow} for each COC:

$\log K_{ow,DCA} = 1.53$, so $K_{ow,DCA} = 34$ L/kg
$\log K_{ow,pyrene} = 4.88$, so $K_{ow,pyrene} = 75,900$ L/kg
$K_{oc,DCA} = 0.63(34) = 22$ L/kg
$K_{oc,pyrene} = 0.63(75,900) = 47,800$ L/kg
$K_{p,DCA} = (0.005)(22) = 0.11$ L/kg
$K_{p,pyrene} = (0.005)(47,800) = 239$ L/kg

Finally, use Eq. 5.23 to find the final concentrations.

$$X_{f,DCA} \approx \frac{500}{1 + \dfrac{3,785}{(889)(0.11)}} = 12.6 \text{ mg}/L$$

$$X_{f,pyrene} \approx \frac{500}{1 + \dfrac{3,785}{(889)(239)}} = 491 \text{ mg}/L$$

The final concentrations in soil would be 12.6 mg/L for DCA and 491 mg/L for pyrene. Pyrene is very hydrophobic, and its K_p value is very high. This example demonstrates that soil washing with water is essentially ineffective in removing pyrene from soil. The addition of surfactants into the washing fluid, using organic solvents, or raising the temperature of the washing fluid, would make soil washing more effective for pyrene.

5.7 OTHER VADOSE ZONE REMEDIATION STRATEGIES

5.7.1 PHYTOREMEDIATION

Phytoremediation is the use of plants to adsorb or degrade COCs. The *in situ* application of phytoremediation involves planting vegetation with roots that reach the contaminated area.

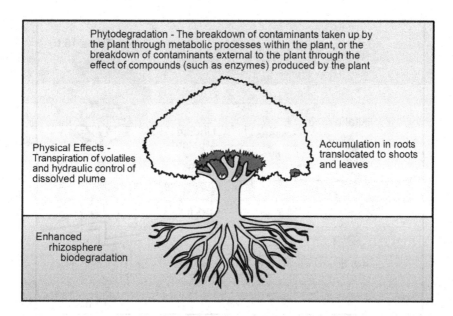

FIGURE 5.14 Examples of mechanisms involved in phytoremediation. (Source: U.S. EPA 2001.)

Mechanisms

Phytoremediation is a general term for the mechanisms described below and shown in Figure 5.14. The points below explain the processes of phytoremediaton.

- Phytoextraction is the uptake and storage of pollutants in the plant's stem or leaves. It is performed by plants called hyperaccumulators, which draw pollutants through their roots. After the pollutants accumulate in the stem and leaves, the plants are harvested and then incinerated. This mechanism is particularly useful for remediating metals.
- Phytovolatilization is a physical mechanism and is the uptake and transpiration of pollutants by a plant. It is performed by plants that take a COC and transform it into an airborne vapor. The vapor can be the pure pollutant, or the pollutant can be metabolized by the plant before it is vaporized.
- Phytodegradation is the metabolization of pollutants by plants, much like biodegradation is the metabolization of pollutants by microorganisms. This mechanism occurs within the plant or in the rhizosphere (rhizodegradation), which is the zone of soil adjacent to the plant roots. After the COC has been drawn into the plant, it assimilates into plant tissues where the plant then degrades the pollutants. The waste products, or daughter compounds, can be volatilized or stored in the plant. If the daughter compounds are relatively benign, the plants can still be used in traditional applications. If the daughter compounds are harmful, then the plants must be properly disposed of.

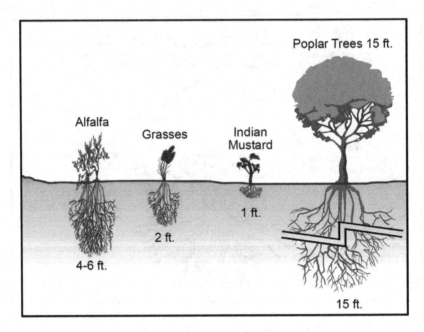

FIGURE 5.15 Examples of typical plants and root depths used in phytoremediation (U.S. EPA 2001.)

Advantages and Disadvantages of Phytoremediation

If a site and its COCs are suitable for phytoremediation, the advantages are many. There is virtually no energy cost because energy is provided by the sun. This technology works for metals and slightly hydrophobic compounds, including many organics. Phytoremediation can stimulate bioremediation in the rhizosphere because plants can stimulate microorganisms through the release of nutrients and the transport of oxygen to the roots. Phytoremediation is relatively inexpensive. Costs include the purchase costs of the plants, planting, monitoring the effectiveness of the technique, and disposing of the plants. Planting vegetation on contaminated sites will reduce erosion by wind and water and leave topsoil intact.

Disadvantages of phytoremediation generally have to do with the limited applicability of this technique. Plants cannot remediate deep aquifers; a tree root can be only as deep as about 15 ft (see Figure 5.15). In addition, it can take many growing seasons to clean up a site. Care must be taken to not cause a different contamination problem when disposing of the contaminated plants. Also, in the case of the phytovolatilization mechanism, care must be taken to not change a soil contamination problem into an air contamination problem.

5.7.2 CAPPING

Remediation does not always need to involve degrading or removing a chemical from a contaminated site. Depending on the risk levels of the contamination, exposure

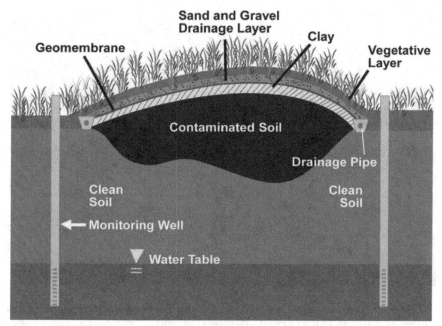

Example of a cover with several layers.

FIGURE 5.16 Capping of contaminated soil using clay and a geomembrane, topped by a drainage layer to remove excess water and vegetation. (Source: U.S EPA 2012.)

pathways, and proximity of receptors, capping is an acceptable remedial solution to isolate and keep COCs in place. Capping consists of covering the area affected by COCs by a clean impermeable surface. Examples of impermeable caps are concrete, asphalt, clay, and geomembranes or impermeable geotextiles (Figure 5.16).

When a site impacted by COCs is capped, two things happen. First, exposure of receptors to COCs is reduced because with an impermeable surface covering the soil, people are prevented from inhaling gases or ingesting contaminated dust. Second, the source of contamination is controlled because with the cap, rainwater cannot infiltrate into the vadose zone and cause the chemical to spread.

Capping is not a "do nothing" alternative and requires periodic inspections and monitoring. Inspections identify whether the caps are cracked or not draining properly. Monitoring means collecting water samples from groundwater monitoring wells to evaluate whether the COCs have become mobile and contaminated groundwater.

5.7.3 SOLIDIFICATION/STABILIZATION (S/S)

Solidification and stabilization are two different technologies often applied together. Solidification is the application of a binding agent (a chemical that renders something solid) to the COC-affected subsurface so that the COC is encapsulated within a solid material. This process may or may not involve a chemical reaction between the

FIGURE 5.17 A schematic of a solidification application where a binding agent is injected into the soil, rendering COCs immobile. (Source: ITRC 2011.)

COC and the binding agent. Stabilization is a chemical process by which the COC becomes less mobile, soluble, or toxic. The combination of solidification and stabilization involves chemical reactors to render the COCs immobile in a solid.

S/S is most often applied as source control of inorganic COCs (U.S. EPA 2007), but it is applicable to a variety of inorganic (heavy metals) and organic (pesticides/herbicides, polycyclic aromatic hydrocarbons (PAHs), volatile organic compounds (VOCs), polychlorinated biphenyls (PCBs), and dioxins/furans (ITRC 2011; U.S. EPA 1993).

S/S can be applied *in situ* or *ex situ*. *In situ* configurations are similar to those shown in Figure 5.17, where the COCs are left in place and an S/S agent is injected into the soil to form a grout-like substance to encapsulate the COCs. When S/S is applied *ex situ*, the COC-impacted soil is excavated and placed in tanks in which S/S agents are mixed. The encapsulated substance is later disposed of.

5.8 SUMMARY

This chapter discussed several technologies to clean up vadose zone soil, which is often the source area for groundwater contamination. Techniques covered were:

- Excavation
- Soil vapor extraction (SVE)
- Bioremediation
- *In situ* chemical oxidation (ISCO)

- Soil washing
- Phytoremediation
- Capping
- Solidification and stabilization

5.9 PROBLEMS AND ACTIVITIES

5.1. A leaking 20 m³ underground storage tank was removed. The excavation resulted in a tank pit of 4 m × 4 m × 5 m, and the excavated soil was stockpiled on site. Three samples were taken from the pile, and the total petroleum hydrocarbon (TPH) concentrations were determined to be ND (<100 mg/kg), 1,500 mg/kg, and 2,000 mg/kg. The bulk density of the soil is 1.8 g/cm³, the fluffy factor is 1.1, and the density of TPH is 0.80 kg/L. What is the mass (in kg) and volume (in L) of TPH in the stockpile?

5.2. You are working on an excavation of contaminated soil and need to estimate the costs of transporting and disposing of the soil. The dimensions of the excavation are 12 m × 14 m × 3 m. The fluffy factor is 1.15, and the *in situ* bulk density of the soil is 1.85 g/cm³.

   a) What is the mass of the stockpiled soil?

   b) Calculate the cost of the disposal of the soil, given the following information. The disposal fee (also known as tipping fee) for the soil at the hazardous waste landfill is the higher of $100/ton or $100/m³. The State charges a tax of $32 per ton of hazardous waste. Each generator of hazardous waste (that is, your consulting firm's client responsible for paying for the remediation of this site) must pay a fee of $28 per ton to the State. Hazardous waste must be transported to the landfill using a hazardous waste manifest (RCRA tracking system) and permitted vehicles. Assume a 275 km trip at $3.75 per km per truck. One truck can haul 40 m³ of waste.

5.3. This problem carefully describes a contamination scenario and should be solved first by sketching the site and reviewing Chapter 4 on how to interpret soil borings.

   A gasoline station contained three 5,000-gallon steel tanks that were excavated and removed. During tank removal it was observed that the tank backfill soil exhibited a strong gasoline odor. Based on visual observations, the fuel hydrocarbon in the soil appeared to have been caused by overspillage of gasoline during filling at unsealed fill boxes or minor piping leakage at the eastern end of the tanks. The excavation resulted in a pit of 30 ft × 20 ft × 18 ft (L × W × H). The excavated soil was stockpiled on site. Four samples were taken from the piles and analyzed for total petroleum hydrocarbons (TPH). The TPH concentrations were ND (not detected, <10 mg/kg), 200, 400, and 800 mg/kg, respectively.

   The tank pit was then backfilled with clean dirt and compacted. Six vertical soil borings (two within the excavated area) were drilled to characterize the subsurface geological condition and to delineate the plume.

The borings were drilled using the hollow-stem-auger method. Soil samples were taken by a 2-inch-diameter split-spoon sampler with brass soil sample retainers every 5 ft below ground surface (bgs). The water table was at 50 ft bgs, and all the borings were terminated at 70 ft bgs. All the borings were then converted to 4-inch groundwater monitoring wells.

Selected soil samples from the borings were analyzed for TPH and benzene, toluene, ethylbenzene, and xylenes (BTEX). The analytical results indicated that the samples from the borings outside the excavated area were all ND. The other results were as listed below:

Boring No.	Depth (ft)	TPH (mg/kg)	Benzene (µg/kg)	Toluene (µg/kg)
B1	25	800	10,000	12,000
B1	35	2,000	25,000	35,000
B1	45	500	5,000	7,500
B2	25	<10	<100	<100
B2	35	1,200	10,000	12,000
B2	45	800	2,000	3,000

The fluffy factor of the soil is 1.15; the porosity of the soil is 0.35, and the bulk density of the soil is 1.8 g/cm³.

Assuming that the leakage contaminated a rectangular block defined by the bottom of the tank pit and the surface of the water table, with length and width equal to those of the tank pit, estimate the following:
a) Total volume of the soil stockpiles (in cubic yards)
b) Mass of TPH in the stockpiles (in kilograms)
c) Volume of the contaminated soil left in the vadose zone (in cubic yards)
d) Mass of TPH, benzene, and toluene in the vadose zone (in kilograms)

5.4. The vadose zone soil in an industrial site is impacted with a chemical solvent. The chemical consists of a mixture of 25% toluene and 75% xylenes by weight. Soil vapor extraction (SVE) is being considered for remediating the site. Estimate the maximum toluene and xylenes concentration of the extracted vapor in mg/m³ (at 1 atm and 20°C), considering a base mass of solvent of 1000 g in the soil.

5.5. Calculate the radius of influence of a vapor extraction well by using vapor drawdown data. The absolute pressure at the extraction well is 0.79 atm; the pressure at a monitoring well 25 ft away is 0.97 atm; and the diameter of the vapor extraction well is 4 inches.

5.6. Calculate the radius of influence of a vapor extraction well by using vapor drawdown data. The absolute pressure at the extraction well is 81 kPa; the pressure at a monitoring well 10 m away is 97 kPa; and the diameter of the vapor extraction well is 4 inches.

5.7. Calculate the radius of influence of a vapor extraction well by using vapor drawdown data. The gauge pressure at the extraction well is −27 in-Hg; the gauge pressure at a monitoring well 25 ft away is −4 in-Hg; and the diameter of the vapor extraction well is 4 inches.

5.8. Calculate the radius of influence of a vapor extraction well by using vapor drawdown data. The absolute pressure at the extraction well is 0.88 atm; the absolute pressure at a monitoring well 40 ft away is 0.98 atm; and the diameter of the vapor extraction well is 4 inches. Consider the radius of influence to be at a location where the pressure is the atmospheric pressure minus 1% of the vacuum drawdown in the extraction well.

5.9. Using the given pressure drawdown data, estimate the pressure (vacuum) in a monitoring well located 18 ft away from the extraction well. The absolute pressure at the extraction well is 0.88 atm; the pressure at a monitoring well 40 ft away is 0.98 atm; the radius of influence is 128 ft at P_{RI} 1 atm; and the diameter of the vapor extraction well is 4 inches. Solve this problem with two different combinations of wells and show that the answer is the same.

5.10. A soil venting well (4 inch diameter) was installed at a site. The pressure in the extraction well is 0.88 atm and the radius of influence of this soil venting well has been determined to be 48 ft. The radial Darcy velocity right outside the well casing was determined as 59 ft/day. Calculate the radial Darcy velocity 22 ft away from the center of the venting well by using Eq. 5.9 (u_r equation).

5.11. Calculate the flow rate entering a soil vapor extraction well given the following information:
- Pressure at the extraction well = 0.85 atm
- Flow rate measured at the extraction pump discharge = 0.21 m³/min

5.12. Oxygen-enhanced biostimulation is being considered at a site during a hot summer. Determine the mass concentration of oxygen in ambient air at 35°C when its volume concentration is approximately 21%. Use the Ideal Gas Law and Eq. 2.5 in Chapter 2.

5.13. For nutrient-enhanced *ex situ* biostimulation, estimate the mass of ammonium sulfate $((NH_4)_2SO_4)$ and tri-sodium phosphate $(Na_3PO_4 \cdot 12H_2O)$ to bioremediate a soil stockpile containing 205 kg of benzene (C_6H_6) and 84 kg of toluene (C_7H_8). A feasibility study has determined that the optimum molar ratio of C:N:P to stimulate microbes to biodegrade the carbon molecules is 100:11:1.5.

5.14. The soil in an excavation is impacted by xylene at a concentration of 5,000 mg/kg. Chemical oxidation is considered as one of the remedial alternatives, with either oxygen or sodium persulfate $(Na_2S_2O_8)$ as the oxidant. Determine the stoichiometric amount of oxidant that needs to be delivered to the impacted zone.

Using oxygen as the oxidant:

$$C_6H_4(CH_3)_2 + 10.5O_2 \rightarrow 8CO_2 + 5H_2O$$

Using sodium persulfate as the oxidant, consider that, as shown in Table 5.3, the stoichiometric requirement of sodium persulfate will be two times that of oxygen.

5.15. A sandy subsurface is contaminated with 300 mg/kg of 1,2-DCA. Soil washing is proposed to remediate the soil. A batch washer that can accommodate 500 kg of soil is designed. For each batch of operation, how much clean water, theoretically, should be added as the washing fluid to reduce the 1,2-DCA concentration to below 10 mg/kg? The bulk density of soil is 1.9 g/cm³ and the fraction of organic carbon of the soil = 0.008.

5.16. A washer uses water to clean soil impacted with chloroethane (C_2H_5Cl). For each batch of washing, 2,000 gallons of water is used for one m³ of soil (bulk density of soil = 1,800 kg/m³). The initial chloroethane concentration in soil is 250 mg/kg and the soil–water partition coefficient (K_p) is 0.4 L/kg. Assuming water and soil reach an equilibrium at the end of washing,

   a) estimate the final concentration of chloroethane in soil in mg/kg
   b) estimate the final concentration of chloroethane in the washing fluid

REFERENCES

Crawford, R., Surbeck, C. Q., Worley, S. B., and Capps, H. Q. P. (2012). "Multiphase Extraction Radius of Influence: Evaluation of Design and Operational Parameters." *Remediation*, 22(4), 37–48.

ITRC. (2005). *Technical and Regulatory Guidance for In Situ Chemical Oxidation of Contaminated Soil and Groundwater*. Interstate Technology & Regulatory Council, Washington, DC.

ITRC. (2011). *Development of Performance Specifications for Solidification/ Stabilization*. Interstate Technology & Regulatory Council, Washington, DC.

Johnson, P. C., Kemblowski, M. W., and Colthart, J. D. (1990a). "Qualitative Analysis for the Cleanup of Hydrocarbon-Contaminated Soils by In-Situ Soil Venting." *Groundwater*, 28(3), 413.

Johnson, P. C., Stanley, C. C., Kemblowski, M. W., Byers, D. L., and Colthart, J. D. (1990b). "A Practical Approach to the Design, Operation, and Monitoring of In Situ Soil-Venting Systems." *Ground Water Monitoring Review*, Spring, 159–178.

Kuo, J. (2014). *Practical Design Calculations for Groundwater and Soil Remediation*. CRC Press. Boca Raton, FL.

Kuo, J. F., Aieta, E. M., and Yang, P. H. (1991). "Three-Dimensional Soil Venting Model and Its Applications." *Emerging Technologies in Hazardous Waste Management II*, D. W. Tedder and F. G. Pohland, eds., American Chemical Society Symposium Series 468, Washington, DC., 382–400.

NIOSH. (2007). *NIOSH Pocket Guide to Chemical Hazards*. DHHS (NIOSH) Publication No. 2005-149. Department of Health and Human Services, Centers for Disease Control and Prevention, National Institute for Occupational Safety and Health, Cincinnati, OH.

Ong, S. K., and Kolz, A. (2007). "Chemical Treatment Technologies." *Remediation Technologies for Soils and Groundwater*, A. Bhandari, R. Surampalli, P. Champagne, S. K. Ong, R. D. Tyagi, and I. Lo, eds., American Society of Civil Engineers, Reston, VA, 79–132.

OSHA. (2011). *Trenching and Excavation Safety*. Occupational Safety and Health Administration, Washington, DC, 2.

Siegrist, R. L., Urynowicz, M. A., West, O. R., Crimi, M. L., and Lowe, K. S. (Eds.). (2001). *Guidance for In Situ Chemical Oxidation at Contaminated Sites: Technology Overview with a Focus on Permanganate Systems*. Battelle Press, Columbus, OH.

U.S. EPA. (1991). *Site Characterization for Subsurface Remediation.* EPA/625/4-91/026. U.S. Environmental Protection Agency Office of Research and Development, Washington, DC.

U.S. EPA. (1993). *Engineering Bulletin Solidification/Stabilization of Organics and Inorganics.* EPA/540/S-92/015. U.S. Environmental Protectio Agency Office of Emergency and Remedial Response, Washington, DC.

U.S. EPA. (1996). *A Citizen's Guide To Soil Washing.* EPA 542-F-96-002. U.S. Environmental Protection Agency Office of Solid Waste and Emergency Response, Cincinnati, OH.

U.S. EPA. (2001). *Brownfields Technology Primer: Selecting and Using Phytoremediation for Site Cleanup.* EPA 542-R-01-006. US Environmental Protection Agency Office of Solid Waste and Emergency Response, Washington, DC.

U.S. EPA. (2002). *Guidance on Choosing a Sampling Design for Environmental Data Collection for Use in Developing a Quality Assurance Project Plan.* EPA/240/R-02-005. U.S. Environmental Protection Agency Office of Environmental Information, Washington, DC.

U.S. EPA. (2007). *Treatment Technologies for Site Cleanup: Annual Status Report 12th Edition.* EPA-542-R-07-012. U.S. Environmental Protection Agency Office of Solid Waste and Emergency Response, Washington, DC.

U.S EPA. (2012). *A Citizen 's Guide To Capping.* EPA 542-F-12-004. U.S. Environmental Protection Agency Office of Solid Waste and Emergency Response, Washington, DC.

U.S. EPA. (2012). *A Citizen's Guide to In Situ Chemical Oxidation.* EPA 542-F-12-011. U.S. Environmental Protection Agency Office of Solid Waste and Emergency Response, Washington, DC.

U.S. EPA. (2017). *Soil Vapor Extraction. How To Evaluate Alternative Cleanup Technologies For Underground Storage Tank Sites.* EPA 510-B-17-003. U.S. Environmental Protection Agency Office of Land and Emergency Management, Washington, DC.

U.S. EPA. (2020). *Town of Pines Groundwater Plume.* https://cumulis.epa.gov/supercpad/SiteProfiles/index.cfm?fuseaction=second.photovideoaudio&id=0508071 (Jan. 30, 2020).

U.S. EPA, Huling, S., and Pivetz, B. (2006). *Engineering Issue: In-situ Chemical Oxidation.* EPA/600/R-06/072. U.S. Environmental Protection Agency Office of Research and Development, Cincinnati, OH.

6 Groundwater Remediation

6.1 INTRODUCTION

The previous chapter addressed remediation in the vadose zone. This chapter introduces groundwater remediation, which is the clean-up of the saturated zone below the vadose zone. While remediating the vadose zone aims to remove the bulk of the mass of contaminants, which are a source of groundwater contamination, remediating the saturated zone aims to reduce contaminant movement and protect sensitive receptors, such as drinking water sources and surface water bodies used for fishing and recreation. In this book, we focus on unconfined aquifers that are located just beneath the vadose zone and are therefore more subject to contamination than the deeper confined aquifers.

Groundwater remediation techniques can be divided into *ex situ* and *in situ*. Pump-and-treat is the *ex situ* technique in which groundwater is extracted from the aquifer and treated aboveground using various technologies. *In situ* techniques involve injecting substances into the aquifer to treat groundwater in place, without groundwater extraction.

6.2 GROUNDWATER PUMPING FOR PUMP-AND-TREAT APPLICATIONS

The most common groundwater remediation technique is pump-and-treat. This technique involves pumping groundwater from strategically located wells, treating the water in a treatment system located aboveground, and discharging the treated water into an appropriate receiving water body (Figure 6.1). Often, the purpose of pump-and-treat is hydraulic containment, which controls the movement of the plume.

6.2.1 GROUNDWATER PUMPING AND EXTRACTION

In Chapter 4, we introduced groundwater monitoring wells, wells used for observation of groundwater flow and quality. In order to extract, or pump, groundwater, extraction wells are installed. When a groundwater extraction well is pumped, the water level (or piezometric surface) in its vicinity declines towards the well. This decline provides an increased hydraulic gradient that drives the water toward the well. The hydraulic gradient gets larger closer to the well.

The decline in groundwater level from the original level is called *drawdown*, as shown in Figure 6.2. An idealized cross-sectional view of the drawdown has the shape of an inverted cone. This results in the so-called *cone of depression*, shown in

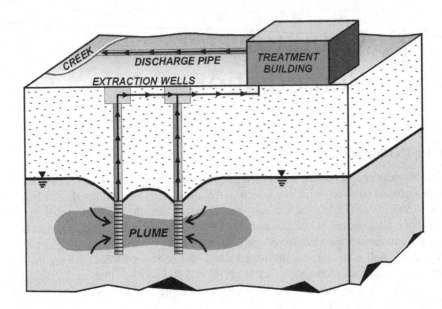

FIGURE 6.1 A schematic of a pump-and-treat system. (Source: U.S. EPA 2001.)

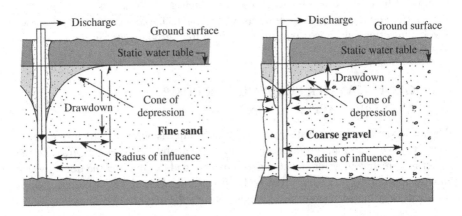

FIGURE 6.2 A schematic of two different unconfined aquifers (fine sand on the left, coarse gravel on the right), the static water table, drawdown, cone of depression, and radius of influence. (Source: Davis and Masten 2013 © McGraw Hill Education.)

Figures 6.2 and 6.3. From a plan view, the radius of the cone of depression represents the water that flows towards the well. This distance is called the *radius of influence*. This is an idealized scenario, and the real cone of depression is not perfectly symmetrical. Figure 6.2 also shows the steepness and radius of influence of the cone of depression according to the soil type. Fine soils with low hydraulic conductivity produce steeper cones of depression and shorter radii of influence. The opposite is true for coarser soils with higher hydraulic conductivity.

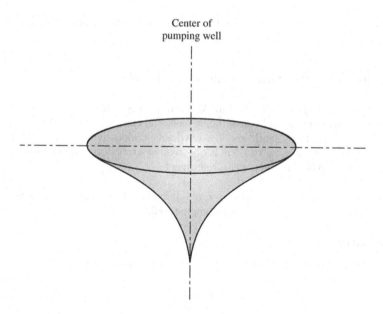

Center of
pumping well

FIGURE 6.3 An idealized cone of depression. (Source: Davis and Masten 2013 © McGraw Hill Education.)

The fundamental equations for flow into wells are described below, first for an unconfined aquifer, which is more commonly contaminated, then for a confined aquifer. For both cases, the fundamental equations were developed with the assumptions that the well completely penetrates the aquifer and that the aquifer is homogeneous and isotropic.

Steady Flow in an Unconfined Aquifer

Under pumping conditions in a fully penetrating extraction well in an unconfined aquifer, the flow rate into the well is described by Equation 6.1.

$$Q = \frac{\pi K \left(h_2^2 - h_1^2 \right)}{\ln \left(\frac{r_2}{r_1} \right)} \tag{6.1}$$

where Q is the flow rate, K is the hydraulic conductivity of the aquifer, h_1 and h_2 are the static heads (water table levels) measured from the bottom of the aquifer (where location 2 is farther from the well than location 1), and r_1 and r_2 are the radial distances from the pumping well, where 1 and 2 are the same locations as for h_1 and h_2.

In an unconfined aquifer, static heads and drawdowns are related in the following way:

$$h + s = B \tag{6.2}$$

where h is the static head (water table level) measured from the bottom of the aquifer, s is the drawdown, and B is the aquifer thickness.

Example 6.1: Steady flow in an unconfined aquifer

An unconfined aquifer is 12.2 m thick. Groundwater is being extracted from a 4-inch (0.1 m) diameter, fully penetrating well. The extraction rate is 0.15 m^3/min. The aquifer is relatively sandy and has a hydraulic conductivity of 1.0×10^{-2} cm/s. Steady-state drawdown, s_1, of 1.5 m is observed in a monitoring well at a distance r_1 of 3.0 m from the pumping well. Determine:

(a) the drawdown s_2 at a distance r_2 of 6.0 m from the pumping well, and
(b) the radius of influence of the pumping well.

SOLUTION:

(a) First determine the static head h_1 (at $r_1 = 3$ m), per Equation 6.2:

$$h = B - s$$

$$h_1 = 12.2 - 1.5 = 10.7 \text{ m}$$

Convert hydraulic conductivity K to units of m/min:

$$K = 1.0 \times 10^{-2} \frac{cm}{s} \times \frac{m}{100 \text{ cm}} \times \frac{60 \text{ s}}{min} = 0.006 \text{ m/min}$$

Use Eq. 6.1 to solve for static head h_2:

$$Q = \frac{\pi K \left(h_2^2 - h_1^2 \right)}{\ln \left(r_2 / r_1 \right)}$$

$$0.15 = \frac{\pi (0.006) \left[h_2^2 - (10.7)^2 \right]}{\ln \left(6 / 3 \right)}$$

$$h_2 = 10.95 \approx 11 \text{ m}$$

Now use Eq. 6.2 again to find the drawdown s_2:

$$s_2 = B - h_2$$

$$s_2 = 12.2 - 10.95 = 1.2 \text{ m}$$

(b) To determine the radius of influence of the pumping well, set the farther r as the radius of influence (r_{RI}) at the location where the drawdown is equal to zero, so h=B. We can use the drawdown information of the pumping well as:

$$0.15 = \frac{\pi(0.006)\left[12.2^2 - 10.7^2\right]}{\ln\left(\frac{r_{RI}}{3}\right)}$$

$$\ln\left(\frac{r_{RI}}{3}\right) = \frac{\pi(0.006)\left[12.2^2 - 10.7^2\right]}{0.15}$$

$$\ln(r_{RI}) - \ln(3) = 4.3165$$

$$\ln(r_{RI}) = 5.4152$$

$$r_{RI} = 225\,m$$

Discussion: For part (b), using either the original r_1 and s_1 or r_2 and s_2 results in the same r_{RI}. This works as long as you keep track of the numbers you input into the farther distance and the closer distance from the pumping well.

Steady Flow in a Confined Aquifer

Steady-state flow from a fully penetrating well in a confined aquifer is described by Equation 6.3.

$$Q = \frac{2\pi KB(h_2 - h_1)}{\ln\left(\frac{r_2}{r_1}\right)} \tag{6.3}$$

Where the variables are as described in Equations 6.1 and 6.2.

In a confined aquifer, static heads and drawdowns are related in the following way:

$$h + s = H \tag{6.4}$$

where h is the static head measured from the bottom of the aquifer, s is the drawdown, and H is the original static head measured from the bottom of the aquifer, which is higher than the aquifer thickness. Figure 6.4 shows the dimensions around a well pumping in a confined aquifer.

Example 6.2: Steady flow in a confined aquifer

A confined aquifer with thickness 9.1 m and hydraulic conductivity 16.3 m/d has a fully-penetrating extraction well. Drawdowns of 0.61 m and 0.36 m are observed in monitoring wells 1.52 m and 6.00 m away. Determine:

(a) the groundwater extraction rate and
(b) the radius of influence of the pumping well.

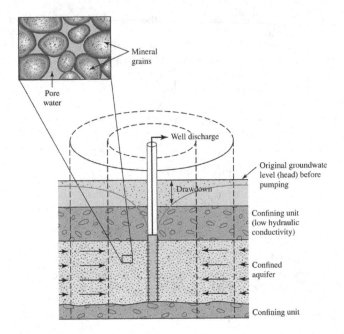

FIGURE 6.4 A pumping well in a confined aquifer. (Source: Davis and Masten 2013 © McGraw Hill Education.)

SOLUTION:

(a) Inserting data in Eq. 6.3, we obtain

$$Q = \frac{2\pi KB(h_2 - h_1)}{\ln\left(\frac{r_2}{r_1}\right)} = \frac{2\pi\left(16.3\frac{m}{d}\right)(9.1\,m)(h_2 - h_1)}{\ln(6.00/1.52)}$$

To solve the term $(h_2 - h_1)$, we can refer to Equation 6.4. We don't know H, but we know it's constant. So we can express h as H − s:

$$h_2 - h_1 = \left[(H - s_2) - (H - s_1)\right] = s_1 - s_2$$

Therefore, we can solve for Q:

$$Q = \frac{2\pi KB(s_1 - s_2)}{\ln\left(\frac{r_2}{r_1}\right)} = \frac{2\pi\left(16.3\frac{m}{d}\right)(9.1\,m)(0.61 - 0.36)}{\ln(6.00/1.52)} = 170\,m^3/d$$

(b) To determine the radius of influence of the pumping well, set r_2 as the radius of influence (r_{RI}) at the location where the drawdown is equal to zero ($s_2 = 0$). In this example, let's use the same r_1 and s_1 as in part (a):

$$170 = \frac{2\pi(16.3)(9.1)[0.61-0]}{\ln\left(\frac{r_{RI}}{1.52}\right)}$$

$$\ln\left(\frac{r_{RI}}{1.52}\right) = \frac{2\pi(16.3)(9.1)(0.61)}{170}$$

$$\ln(r_{RI}) - \ln(1.52) = 3.344$$

$$\ln(r_{RI}) = 3.763$$

$$r_{RI} = 43 \text{ m}$$

DISCUSSION:

1. Substituting $h_2 - h_1$ for $s_1 - s_2$ only works for Eq. 6.3 for confined aquifers. This substitution does not work for Eq. 6.1 for unconfined aquifers because h_2 and h_1 are squared.
2. For part (b), using either the original r_1 and s_1 or r_2 and s_2 results in the same r_{RI}. This works as long as you keep track of the numbers you input into the farther distance and the closer distance.

6.2.2 CAPTURE ZONE ANALYSIS

Equations 6.1 and 6.3 are not sufficient for a full analysis of how much groundwater is captured in a pumping well. Because groundwater is flowing, the idealized cone of depression around an extraction well is only realistic to some extent. The purpose of a capture zone analysis is to determine where to locate extraction wells so that they capture the entire plume. Extraction wells should be strategically located to create a capture zone that encloses the entire plume and does not let any COCs escape.

One Groundwater Extraction Well

Figure 6.5 shows a plan view of groundwater moving in the x-direction, from right to left. A groundwater extraction well is located at the intersection of the x- and y-axes. The curve around the well and elongated toward the positive x-axis is called the capture zone envelope, which encompasses the groundwater captured by the extraction well.

The capture zone envelope is defined by Equation 6.5, where y is the dimension of the capture zone envelope in the positive or negative direction of the y-axis. This analysis assumes a homogeneous and isotropic aquifer with a uniform thickness and that the groundwater flow is uniform and steady (Javandel and Tsang 1986).

$$y = \pm \frac{Q}{2Bu} - \frac{Q}{2\pi Bu} \tan^{-1} \frac{y}{x} \tag{6.5}$$

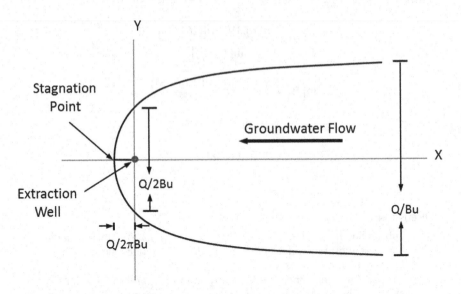

FIGURE 6.5 Capture zone of one extraction well. (Source: Kuo 2014 © Taylor & Francis.)

Where B is the aquifer thickness, Q is the groundwater extraction rate, and u is the Darcy velocity. The Darcy velocity is equal to Ki, where i is the regional hydraulic gradient. y/x is in radians.

From Figure 6.5, we can see that the larger the Q/Bu value, the larger the capture zone is. Three dimensions are important in this analysis:

1. The stagnation point x_{sp}, where y approaches zero. This is the farthest downstream distance that the pumping well can reach.

$$x_{sp} = \frac{Q}{2\pi Bu} \tag{6.6}$$

2. The side-stream distances at the line of the extraction well, where $x = 0$.

$$y_{x=0} = \pm \frac{Q}{4Bu} \tag{6.7}$$

3. The asymptotic value of y, where x approaches infinity.

$$y_{x \to \infty} = \pm \frac{Q}{2Bu} \tag{6.8}$$

In Eqs. 6.7 and 6.8, as the ± signs indicate, y is the width of the capture zone in one direction of the y-axis only, so the total width of the plume is twice that magnitude, or Q/2Bu and Q/Bu, respectively. Eqs. 6.6 through 6.8 give an estimate of the rough shape of the capture zone. It is straightforward to plot x_{sp} and $y_{x=0}$. However, plotting $y_{x \to \infty}$ is not possible because of the infinite value of x. A more practical dimension to plot the capture zone far away from the extraction well is the distance $x = 10y_{x=0}$. The width y at $x = 10y_{x=0}$ can then be calculated using Eq. 6.5, by trial-and-error or using a solver function in a calculator or spreadsheet.

To obtain a more detailed shape of the capture zone envelope, Eq. 6.5 can be solved for x and rearranged as shown in Eqs. 6.9 and 6.10.

$$x = \frac{y}{\tan\left\{\left[+1 - \left(\frac{2Bu}{Q}\right)y\right]\pi\right\}} \quad \text{for positive y values} \tag{6.9}$$

$$x = \frac{y}{\tan\left\{\left[-1 - \left(\frac{2Bu}{Q}\right)y\right]\pi\right\}} \quad \text{for negative y values} \tag{6.10}$$

A set of (x, y) coordinates can be obtained from these equations by setting values for y and solving the equation for x. The envelope is symmetrical about the x-axis.

Example 6.3: Draw the Envelope of a Capture Zone of a Groundwater Pumping Well

Calculate x_{sp}, $y_{x=0}$, $y_{x\to\infty}$, y at $x = 10y_{x=0}$, and delineate the capture zone of a groundwater recovery well with the following information:

- $Q = 8.0$ ft³/min
- Hydraulic conductivity = 267 ft/day
- Groundwater gradient = 0.01
- Aquifer thickness = 50 ft

SOLUTION:

(a) Determine the groundwater velocity, u:

$$u = (K)(i) = \left[(267 \text{ ft/day})(1 \text{ d/1,440 min})\right](0.01)$$

$$= 1.85 \times 10^{-3} \text{ ft/min}$$

The stagnation point is

$$x_{sp} = \frac{Q}{2\pi Bu} = \frac{8.0 \text{ ft}^3/\text{min}}{2\pi(50 \text{ ft})(1.85 \times 10^{-3} \text{ ft/min})} = 13.8 \text{ ft}$$

(b) The side-stream distance at the well is

$$y_{x=0} = \pm\frac{Q}{4Bu} = \pm\frac{8.0}{4(50 \text{ ft})(1.85 \times 10^{-3} \text{ ft/min})} = 21.6 \text{ ft}$$

And the full-width side-stream distance is 2×21.6 ft = 43.2 ft

(c) The width of the plume when x approaches infinity is

$$y_{x\to\infty} = \pm\frac{Q}{2Bu} = \pm\frac{8.0}{2(50\,\text{ft})(1.85\times10^{-3}\,\text{ft}/\text{min})} = 43.2\,\text{ft}$$

And the full-width as x approaches infinity is $2\times43.2\,\text{ft}=86.4\,\text{ft}$
(d) To calculate y at $x=10y_{x=0}$, first calculate $10y_{x=0}$:

$$x = 10\times y_{x=0} = 10(21.6\,\text{ft}) = 216\,\text{ft}$$

Then use Eq. 6.5 and solve by trial-and-error or using a calculator or spreadsheet solver function, remembering that y/x is in radians:

$$y = \pm\frac{Q}{2Bu} - \frac{Q}{2\pi Bu}\tan^{-1}\frac{y}{x}$$

$$y = \pm\frac{8.0}{2(50\,\text{ft})(1.85\times10^{-3}\,\text{ft/min})} - \frac{8.0}{2\pi(50\,\text{ft})(1.85\times10^{-3}\,\text{ft/min})}\tan^{-1}\frac{y}{216}$$

Using a spreadsheet, $y=\pm40.6\,\text{ft}$, so the full-width is 81.2 ft
(e) To delineate the capture zone, establish a set of (x, y) values using Eqs. 6.9 and 6.10. First, specify values of y (select smaller intervals for small y values). The minimum y value should be zero, and the maximum y value can be approximately 40.6 ft, which was found in part (d). The following table lists some of the data points used to plot Figure 6.6.

y (ft)	x (ft)
0.0	0
0.1	−13.8
1	−13.7
5	−13.1
10	−11.2
20	−2.34
30	21.0
40	169
−0.1	−13.8
−1	−13.7
−5	−13.1
−10	−11.2
−20	−2.34
−30	21.0
−40	169

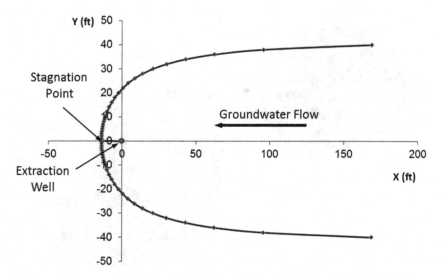

FIGURE 6.6 Capture zone of the extraction well (Example 6.3). (Source: Kuo 2014 © Taylor & Francis.)

DISCUSSION:

1. The capture zone curve is symmetrical about the x-axis, as shown in the table and the figure. Note that Eq. 6.9 should be used for positive y values and Eq. 6.10 for negative y values.
2. Solving Eq. 6.9 or 6.10 for x using a very small value for y (but not zero) results in the stagnation point.
3. For the capture zone delineation, do not specify the y values beyond the values of $\pm Q/2Bu$. As discussed, $\pm Q/2Bu$ are the asymptotic values of the capture zone curve ($x=\infty$).
4. The y value when $x=10y_{x=0}$ is a good approximation for $y_{x\to\infty}$.

6.3   *EX SITU* GROUNDWATER TREATMENT TECHNOLOGIES

Many technologies exist for treating water when a pump-and-treat strategy is used. In Section 6.2, groundwater pumping was addressed. In this section, we examine the technologies for treating extracted groundwater. These are considered to be *ex situ* technologies because the groundwater is removed from its location.

6.3.1   ACTIVATED CARBON ADSORPTION

Adsorption is the process that collects soluble substances (adsorbates) in solution onto the surface of the solid (the adsorbent). This term is not to be confused with absorption. In absorption, the absorbate moves into the bulk of the absorbent, much

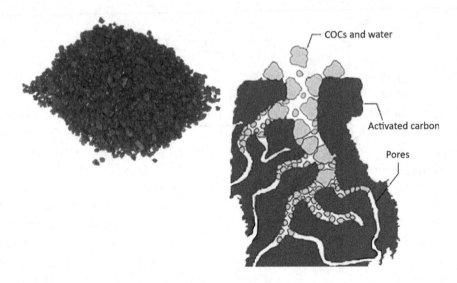

FIGURE 6.7 Left: Granular activated carbon (GAC). Right: Schematic of a partial activated carbon grain with tree-root structure pores. (Modified from U.S. EPA 2012a.)

as water moves into a sponge. In adsorption, adsorbates are adhered, or stuck, onto the adsorbent. Soil is one type of adsorbent, as seen in Section 2.5.5. When applied to groundwater remediation, the adsorbate is the COC, the solution is water, and the adsorbent is often the activated carbon. Activated carbon is a universal adsorbent that adsorbs almost all types of organic compounds. It is a substance similar to charcoal (see Figure 6.7) but manufactured specifically to be an adsorbent. Granular activated carbon (GAC) is very porous (see Figure 6.7) and therefore has a large specific surface area (surface area per unit mass of the grain). GAC is packed inside specially designed vessels, or tanks, through which pumped groundwater flows (Figure 6.8). The combination of GAC inside a vessel is called a unit. When the water flows through the units, the organic compounds leave the extracted groundwater by adsorbing onto the GAC surface. As the COC molecules continue to move through the unit, they take up the surface area of the GAC. When the surface area of the GAC becomes saturated, the unit is said to be exhausted, as indicated by the breakthrough of COCs in the effluent. The GAC in the unit then needs to be regenerated or replaced.

Common preliminary design of a GAC adsorption system includes sizing the adsorption unit, determining the carbon change (or regeneration) interval, and configuring the carbon units, if multiple adsorption units are used.

Adsorption Isotherm and Adsorption Capacity

In general, the extent of adsorption depends on the characteristics of the adsorbates (i.e., the COCs) and the GAC, the concentrations of the COCs, and the temperature. An adsorption isotherm (from Greek, *iso* meaning *equal*, and *therm* meaning *heat*) describes the equilibrium relationship between the adsorbed COC concentration on

FIGURE 6.8 GAC vessels with connecting piping. (Photo credit: Michael Shiang.)

the surface of the GAC and the dissolved COC concentration in the solution at a constant temperature. The adsorption capacity of a given GAC for a specific compound is estimated from their isotherm data. In practice, these data are developed by the manufacturer. Mathematical adsorption models can be used to predict the extent of adsorption. The most commonly used adsorption models in environmental applications are the Langmuir and Freundlich isotherms, respectively:

$$q = \frac{abC}{1 + bC} \tag{6.11}$$

$$q = KC^{1/n} \tag{6.12}$$

where q is the adsorbed COC concentration (in mass of COC/mass of activated carbon), C is the influent aqueous COC concentration (in mass of COC/volume of water), and a, b, K, and 1/n are empirical constants.

The adsorbed COC concentration (q) obtained from Eqs. 6.11 and 6.12 is an equilibrium value (the one in equilibrium with the aqueous COC concentration). It should be considered as the theoretical adsorption capacity for a specified aqueous COC concentration because it is developed in a laboratory setting. The actual adsorption capacity in field applications would be lower because of the presence of other compounds that would compete for the adsorption sites. Typically, design engineers take about 50% of this theoretical value as the design adsorption capacity, where 50% is a factor of safety. Therefore,

$$q_{design} = (50\%)(q_{theoretical})$$ (6.13)

The maximum mass of COCs that can be removed from the groundwater and held by a given mass of GAC ($M_{removed}$) can be determined as:

$$M_{removed} = (q_{design})(M_{GAC}) = (q_{design})\left[(V_{GAC})(\rho_b)\right]$$ (6.14)

where M_{GAC} is the mass, V_{GAC} is the volume, and ρ_b is the bulk density of the GAC, respectively. The bulk density of GAC is typically 30 lb/ft^3 or 480 kg/m^3.

When referring to a mass of COCs that flows through a system, it is useful to know the definition of a mass flow rate:

$$\dot{M} = CQ$$ (6.15)

where $\dot{M}$ is the mass flow rate of chemical (mass/time) and Q is the volumetric flow rate (volume/time) of a fluid (water or air). Q=V/t, where V is the volume of a fluid passing through a location within a period of time, and t is the time during which a volume of fluid is passing.

Example 6.4: Determine the Mass of Activated Carbon Used Daily

Extracted groundwater contaminated with 75 µg/L of methylene chloride (MC) flowing at 54 m³/day is to be treated using GAC. Use the Freundlich isotherm to calculate the mass (kg) of GAC used per day. The Freundlich parameters developed for MC are K=6.25 and 1/n=0.801, resulting in a q in units of µg MC/g GAC.

SOLUTION:

(a) The theoretical adsorption capacity can be found by using the Freundlich adsorption isotherm as:

$$q = KC^{1/n} = 6.25(75)^{0.801} = 198 \frac{\mu gMC}{gGAC}$$

The actual adsorption capacity can be found by using Eq. 6.13 as:

$$q_{design} = (50\%)(198) = 99.2 \frac{\mu gMC}{gGAC}$$

(b) The mass of methylene chloride flowing through the GAC every day is found using Eq. 6.15:

$$\dot{M}_{MC} = C_{MC}Q = \left(75\frac{\mu g}{L}\right)\left(54\frac{m^3}{day}\right)\left(\frac{1000\,L}{m^3}\right) = 4.05\times10^6\,\frac{\mu g}{day}$$

(c) Calculate the mass of GAC using q_{design} and the masses of MC and GAC on a daily basis. Remember that q is the adsorbed COC concentration (in mass of COC/mass of activated carbon):

$$q_{design} = \frac{Mass_{MC}}{Mass_{GAC}} = 99.2\frac{\mu g\,MC}{g\,GAC}$$

$$Mass_{GAC} = \frac{Mass_{MC}}{q_{design}} = \frac{4.05\times10^6\,\mu g\,MC}{99.2\frac{\mu g\,MC}{g\,GAC}} = 4.08\times10^4\,g\,GAC = 41\,kg\,GAC$$

DISCUSSION:

1. Care should be taken to use the correct units for C and q in the isotherm equations. Those units are usually found in manufacturer-supplied tables.
2. The influent aqueous COC concentration, not the effluent concentration, should be used in the isotherms to estimate the adsorption capacity.

The following procedure can be used to determine the adsorption capacity of a GAC adsorption unit:

Step 1: Determine the theoretical adsorption capacity of the GAC by using Eq. 6.11 or 6.12.
Step 2: Determine the design adsorption capacity of the GAC by using Eq. 6.13.
Step 3: Determine the mass of GAC in the adsorber.
Step 4: Determine the maximum mass of COCs that can be held by the GAC in the adsorber using Eq. 6.14.

Example 6.5: Determine the Capacity of an Activated Carbon Adsorber

Often during construction activities, the groundwater level is high and interferes with the building foundation. Pumping groundwater to lower its level is called dewatering, and it is necessary for below-ground construction.

At a construction site, the contractor unexpectedly found that the extracted groundwater contained 5.0 mg/L toluene. The toluene concentration of groundwater has to be reduced to below 100 ppb before discharge. To avoid further delay of the tight construction schedule, off-the-shelf 55-gallon activated carbon units are proposed to treat the extracted groundwater.

The activated carbon vendor provided the Langmuir adsorption isotherm information:

$$q\left(\frac{kg\ toluene}{kg\ GAC}\right) = \frac{0.004\ C_{in}}{1+0.002\ C_{in}}$$

where C_{in} is in mg/L. The vendor also provided the following information regarding the adsorber:

- diameter of carbon packing bed in each 55-gallon drum = 1.5 ft
- height of carbon packing bed in each 55-gallon drum = 3 ft
- bulk density of the activated carbon = 30 lb/ft³.

Determine (a) the adsorption capacity of the activated carbon, (b) the volume and mass of activated carbon in each 55-gallon unit, and (c) the mass of the toluene that each unit can remove before exhaustion.

SOLUTION:

(a) The theoretical adsorption capacity can be found by using the given adsorption isotherm as:

$$q\left(\frac{kg\ toluene}{kg\ GAC}\right) = \frac{0.004(5)}{1+0.002(5)} = 0.02\ kg/kg$$

The actual adsorption capacity can be found by using Eq. 6.13 as:

$$q_{design} = (50\%)(0.02) = 0.01\frac{kg\ toluene}{kg\ activated\ carbon}$$

We can consider the number above to be a ratio that can work with any units of mass, for example, 0.01 g toluene/g activated carbon or 0.01 lb toluene/lb activated carbon.

(b) The volume and mass of the GAC inside a 55-gallon drum can be calculated starting with the volume of a cylinder, $\pi r^2 h$:

$$V = \pi\left(\frac{1.5}{2}\right)^2 (3) = 5.3\ ft^3$$

The mass is equal to:

$$Mass_{GAC} = V\rho_b = 5.3(30) = 159\ lb$$

(c) The mass of toluene that can be retained in a drum before the carbon becomes exhausted is the mass removed from the groundwater:

$$M_{removed} = (q_{design})(M_{GAC}) = \left(0.01\frac{lb\ toluene}{lbGAC}\right)(159\ lbGAC) = 1.59\ lbs\ toluene$$

DISCUSSION:

1. The mass of GAC is always much higher than the mass of COC removed.
2. The adsorption capacity of 0.01 kg/kg is equal to 0.01 lb/lb or 0.01 g/g.

3. Care should be taken to use appropriate units for C and q in the isotherm equations, which are dictated by the empirically derived constants.
4. The influent aqueous COC concentration, not the effluent concentration, should be used in the isotherms to estimate the adsorption capacity.

Design of an Activated Carbon Adsorption System

Two parameters are used to design the volume and dimensions of a carbon adsorption unit, the empty bed contact time (EBCT) and surface loading rate (SLR).

Empty Bed Contact Time (EBCT) To design the size of the liquid-phase GAC system, a standard criterion used is the EBCT, which is the nominal time it would take for the water to flow through the GAC bed if it were empty of activated carbon. Typical EBCTs range from 3 to 35 minutes (Ong and Kolz 2007), mainly depending on the characteristics of the COCs. Some compounds have a stronger tendency to adsorb, and the required EBCT would be shorter. For example, polychlorinated biphenyls (PCBs) are very hydrophobic and strongly adsorb to the GAC surface, while acetone is very soluble in water and not readily adsorbable. The required EBCT for acetone would be much longer than that for PCBs.

If the water flow rate (Q) is specified, the EBCT can be used to determine the required volume of the GAC (V_{GAC}) as:

$$V_{GAC} = (EBCT)(Q) \tag{6.16}$$

Surface Loading Rate (SLR) The typical SLR, or hydraulic loading rate, of carbon adsorbers is typically between 1.9 to 30 m^3 h^{-1} m^{-2} (0.8 to 12 gpm/ft^2) (Ong and Kolz 2007). This parameter can be used to determine the minimum required cross-sectional area of the adsorber (A_{GAC}):

$$A_{GAC} = \frac{Q}{SLR} \tag{6.17}$$

After fulfilling the EBCT and SLR requirements, the height of the GAC bed (H_{GAC}) can then be determined as:

$$H_{GAC} = \frac{V_{GAC}}{A_{GAC}} \tag{6.18}$$

COC Removal Rate by the Activated Carbon Adsorber

The COC removal rate from water by a GAC adsorber ($R_{removal}$) can be calculated by using the following mass flow rate formula, where the COC concentration is the difference between the inlet and outlet concentrations:

$$R_{removal} = (C_{in} - C_{out})Q \tag{6.19}$$

In practical applications, the effluent concentration (C_{out}) is kept below the discharge limit, which is often very low. Therefore, for a factor of safety, the term of C_{out} can be deleted from Eq. 6.19 in design. The mass removal rate by the GAC is then essentially the same as the mass loading rate ($R_{loading}$):

$$R_{removal} \cong R_{loading} = C_{in} Q \qquad (6.20)$$

Change-out (or Regeneration) Frequency

Once the activated carbon reaches its capacity, it should be regenerated or disposed of. The expected service life of a fresh batch of activated carbon can be estimated by dividing the capacity of the activated carbon in the adsorber with the COC removal rate ($R_{removal}$) as:

$$T = \frac{M_{removed}}{R_{removal}} \qquad (6.21)$$

Configuration of the Activated Carbon Adsorbers

Activated carbon adsorbers are always configured as multiple units. Usually, at least two activated carbon adsorbers are used in series (Figure 6.9). When two adsorbers are arranged in series, a monitoring point should be located at the effluent of the first adsorber. This is because a high effluent concentration from the first

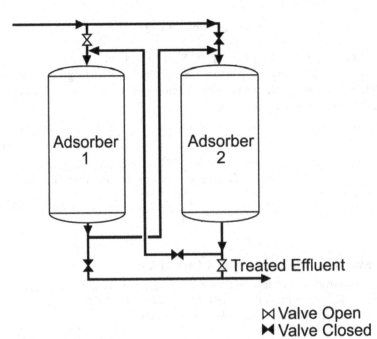

FIGURE 6.9 Two activated carbon adsorbers configured in series. As shown, Adsorber 1 is the primary adsorber (water flows through the open valves). When Adsorber 1 is saturated, the carbon inside the vessel is removed and replaced with fresh carbon. The valves are then opened and closed so that Adsorber 2 becomes the primary adsorber. (Modified from U.S. EPA 2006.)

adsorber indicates that this adsorber is reaching its capacity. The first adsorber is then taken off-line soon, and the second adsorber is shifted to be the first adsorber. Consequently, the capacity of both adsorbers would be fully utilized while meeting environmental permit requirements. Sometimes, more than two activated carbon adsorbers are used, arranged in series and/or in parallel. If there are two parallel streams of adsorbers, one stream can be taken off-line for regeneration or maintenance, while the continuous operation of the process is not interrupted.

The following procedure can be used to complete the design of an adsorption system:

Step 1: Determine the design adsorption capacity as described earlier (also see Ex. 6.4).
Step 2: Determine the required volume of the GAC by using Eq. 6.16.
Step 3: Determine the required area of the activated carbon adsorber by using Eq. 6.17.
Step 4: Determine the required height of the GAC by using Eq. 6.18.
Step 5: Determine the COC removal rate or loading rate by using Eq. 6.20.
Step 6: Determine the mass of the COCs that the GAC adsorber(s) can hold by using Eq. 6.4.
Step 7: Determine the service life of the GAC adsorber by using Eq. 6.21.
Step 8: Determine the optimal configuration when multiple adsorbers are used.

Example 6.6: Design the Configuration of an Activated Carbon System for Groundwater Remediation

The site described in Example 6.5 will use off-the-shelf 55-gallon drums of GAC to remediate the contaminated groundwater. Design an activated carbon treatment system using the information below. The design should include (a) the required volume and cross-sectional area of GAC; (b) the number of GAC units and the configuration of flow; and (c) the GAC change-out frequency.

- required EBCT ≥ 12 minutes
- required SLR ≤ 5 gpm/ft²
- extracted groundwater flow rate = 30 gpm
- diameter of GAC packing bed in each 55-gallon drum = 1.5 ft
- height of GAC packing bed in each 55-gallon drum = 3 ft
- bulk density of GAC = 30 lb/ft³
- adsorption isotherm as in Example 6.5:

$$q_{design} = 0.01 \frac{lb\, toluene}{lb\, GAC}$$

SOLUTION:

(a) The required volume of GAC can be found by using Eq. 6.16:

$$V_{GAC} = (EBCT)(Q) = (12\, min)\left(30\, \frac{gal}{min}\right)\left(\frac{ft^3}{7.48\, gal}\right) = 48.1\, ft^3$$

The required cross-sectional area can be found by using Eq. 6.17:

$$A_{GAC} = \frac{Q}{SLR} = \frac{30 \text{ gpm}}{5 \text{ gpm}/\text{ft}^2} = 6 \text{ ft}^2$$

If the adsorption system were tailor-made, then a system with a cross-sectional area of 6 ft² and a height of 8 ft (=48.1/6) would do the job. However, the contractor has decided to use off-the-shelf 55-gallon drums of GAC, so we need to determine the number of drums that will provide the required cross-sectional area and volume.

(b) To find the number of drums (units), first determine the dimensions of one 55-gallon drum.

Cross-sectional area and volume of the activated carbon inside a 55-gallon drum:

$$A_{GAC\ drum} = \frac{\pi D^2}{4} = \frac{\pi (1.5)^2}{4} = 1.77 \text{ ft}^2 \left(\text{per drum}\right)$$

$$V_{GAC\ drum} = A_{GAC\ drum} H_{GAC\ drum} = 1.77 \times 3 = 5.3 \text{ ft}^3 \left(\text{per drum}\right)$$

Number of drums in parallel to meet the required SLR:

$$N_{drums\ parallel} = \frac{6 \text{ ft}^2}{1.77 \text{ ft}^2/\text{drum}} = 3.4 \text{ drums}$$

So, use four drums in parallel. The total cross-sectional area of four drums in parallel is

$$A_{parallel} = 1.77 \times 4 = 7.07 \text{ ft}^2$$

So far, we have met the SRL requirement. Now we need to meet the EBCT, which is related to the volume and height of GAC.

The required height of the GAC adsorber can be found by using Eq. 6.18:

$$H_{GAC} = \frac{V_{GAC}}{A_{parallel}} = \frac{48.1}{7.07} = 6.8 \text{ ft}$$

The height of activated carbon in each drum is only 3 ft. The number of drums in series to meet the required height of 6.8 ft is:

$$N_{drums\ in\ series} = \frac{H_{GAC}}{H_{drum}} = \frac{6.8}{3} = 2.3 \text{ drums}$$

So, use three drums in series for each of the four parallel trains, for a total of 12 GAC units (Figure 6.10).

The volume of GAC in each parallel train of three drums in series is:

$$V_{series} = V_{drum} \times 3 \text{ drums} = 5.3 \frac{\text{ft}^3}{\text{drum}} \times 3 \text{ drums} = 15.9 \text{ ft}^3$$

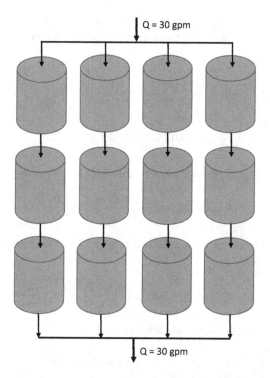

FIGURE 6.10 A configuration of activated carbon vessels using four parallel trains with three vessels in series. The total flow of 30 gpm is split among the four trains (Example 6.6).

Therefore, the total volume of GAC in this system is

$$V_{GAC\ total} = 15.9\ ft^3 \times 4 = 63.6\ ft^3$$

We can double-check that the SLR and EBCT are met, now that we know we have 4 parallel trains of 3 drums each.

$$SLR = \frac{Q}{A_{total}} = \frac{30\ gpm}{7.07\ ft^2} = 4.24\ gpm/ft^2$$

The four parallel trains have an SLR of 4.24 gpm/ft², which is less than the maximum allowed of 5 gpm/ft². Therefore, the SLR requirement is met. The EBCT requirement can also be checked, keeping in mind that we have to divide the total flow rate by four parallel trains:

$$EBCT = \frac{V_{series}}{Q_{series}} = \frac{15.9\ ft^3}{(30\ gpm)/4} \times 7.48\ \frac{gal}{ft^3} = 15.8\ min$$

The EBCT requirement is met.
(c) To determine how often a GAC drum needs to be changed out (the change-out frequency), we need to consider the isotherm. First,

determine the rate at which toluene-laden groundwater is removed from the subsurface (the removal rate, which is a loading rate) by using Eq. 6.20:

$$R_{removal} = \dot{M} = (C_{in})(Q) = \left(5\frac{mg}{L}\right)\left(30\frac{gal}{minute}\right)\left(3.785\frac{L}{gal}\right) = 568\frac{mg}{minute}$$

$$= 0.00125\frac{lb}{minute} = 1.8\frac{lb}{day}$$

Now we must remember that this rate is split into 4 parallel trains:

$$R_{removal\ parallel} = \frac{1.8\ lb/day}{4} = 0.45\frac{lb}{day}$$

Now determine the mass of toluene that each GAC adsorber can hold by using Eq. 6.14:

$$M_{removed} = (q_{design})\left[(V_{GAC})(\rho_b)\right] = 0.01\frac{lb\ toluene}{lb\ GAC}\left[\left(5.3\frac{ft^3}{drum}\right)\left(30\frac{lb}{ft^3}\right)\right]$$

$$= 1.59\ lb\ toluene\ in\ each\ drum$$

The service life of each GAC adsorber, or the time each adsorber can be in service, is found by using Eq. 6.21:

$$T = \frac{M_{removed}}{R_{removal}} = \frac{1.59\ lb/drum}{0.45\ lb/day} = 3.5\frac{days}{drum}$$

This means that the first GAC unit of each train will be changed out after 3.5 days. At this change-out time, the second set of drums in series becomes the primary adsorbers, which then will also be changed out after another 3.5 days, and so on.

In conclusion, we have determined that the minimum volume and cross-sectional area of GAC are 48.4 ft³ and 6 ft², respectively. We have met those requirements by selecting a GAC adsorption system of 12 55-gallon drum units, with a total volume and cross-sectional area of 64 ft³ and 7.1 ft², respectively. The configuration is four parallel trains, each with three drums in series. The estimated service life of each of the four primary drums is 3.5 days.

DISCUSSION:

1. The configuration is 4 parallel trains, each with 3 drums in series (a total of 12 drums). Care should be taken to minimize the head loss resulting from three drums in series and numerous piping connections.
2. A 55-gallon activated carbon drum normally costs several hundred dollars. In this example, four drums will have to be changed out every 3.5 days. The disposal or regeneration cost should also be considered, and that makes this option relatively expensive. If long-term treatment

is needed, one may want to switch to larger GAC adsorbers or other treatment methods.

In conclusion, activated carbon adsorption is one of the most common groundwater remediation techniques for removing organic compounds. The design for activated carbon units depends on isotherm equations developed by the carbon manufacturers for different chemicals, plus recommended surface loading rates (SLRs) and empty bed contact times (EBCTs). Carbon adsorption is considered to be a separation process. That is, it's a method that separates the COC from the fluid. Because of this, the COC accumulates on the adsorption material. Therefore, part of designing and operating activated carbon systems is determining how often the carbon needs to be changed out or regenerated.

6.3.2 AIR STRIPPING

Air stripping is another separation process where the COCs are separated from the water. This is a physical process in which air is mixed with the water to promote the volatilization of the COCs. Air stripping works only with volatile organic compounds (VOCs). As a separation process, air stripping is also considered a mass transfer process, that is, the mass of COCs is transferred from water to air.

The primary type of air stripping system is a packed-column tower. A packed-column air stripper is a cylindrical vessel filled with packing, often a plastic material designed to enable both the water and the air streams to separate into smaller streams, enhancing the contact between water and air. Water is distributed over the top of the packing, while air is blown from the bottom of the tower in an upward motion (Figure 6.11). This process then facilitates the mass transfer of COCs from water to air.

Design of a Packed-Column Air Stripping Tower

In a packed-column air stripping tower, the air and the impacted groundwater streams flow in opposite directions through a packing column. The packing provides a large surface area for the VOCs to migrate from the liquid stream to the air stream. These streams are represented in Figure 6.12 and are further explained using the concept of mass balance.

A mass balance equation can be derived by making the mass of COCs removed from the liquid equal to the mass of COCs entering the air.

$$Q_w\left(C_{in} - C_{out}\right) = Q_a\left(G_{out} - G_{in}\right) \tag{6.22}$$

where C is the COC concentration in the water (mg/L); G is the COC concentration in the air (mg/L); Q_a is the air flow rate (L/min); and Q_w is the water flow rate (L/min)

For an ideal case where the influent air is clean and contains no COCs ($G_{in}=0$) and the groundwater is completely decontaminated ($C_{out}=0$), then Eq. 6.22 is simplified to:

$$Q_w\left(C_{in}\right) = Q_a\left(G_{out}\right) \tag{6.23}$$

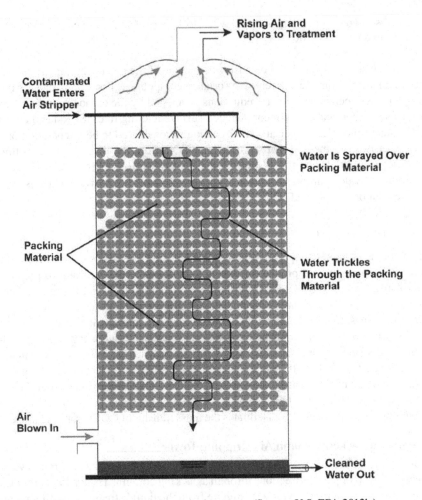

FIGURE 6.11   A packed-column air stripping tower. (Source: U.S. EPA 2012b.)

Because these are volatile chemicals, then Henry's law applies. Assuming the efflu-ent air is in equilibrium with the influent water, then Eq. 2.16 in Chapter 2 becomes:

$$G_{out} = H^*C_{in} \tag{6.24}$$

where H* is the dimensionless Henry's constant of the COC. Combining Equations 6.23 and 6.24:

$$H^*\left(\frac{Q_a}{Q_w}\right)_{min} = 1 \tag{6.25}$$

The Q_a/Q_w ratio is the minimum required air-to-water ratio in an air stripper for the complete volatilization of the COC. In practice, the actual air-to-water ratio is often several times this minimum value that is based on ideal conditions.

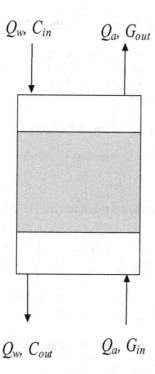

Q_w, C_{in} Q_a, G_{out}

Q_w, C_{out} Q_a, G_{in}

FIGURE 6.12 Streams of water and air into an air stripper. (Source: Kuo 2014 © Taylor & Francis.)

The stripping factor (S) is a factor used in design that is based on the dimensionless Henry's constant and the air-to-water ratio:

$$S = H^* \left(\frac{Q_a}{Q_w} \right) \tag{6.26}$$

S = 1 for the idealized condition. For field applications, S should be greater than 1 and normally ranges from 2 to 10.

The following procedure can be used to determine the air flow rate for a given liquid flow rate:

Step 1: Convert the Henry's constant to its dimensionless value using the formula given in Table 2.3.

Step 2: If the stripping factor is known or selected, determine the air-to-water ratio by using Eq. 6.26. Go to Step 4.

Step 3: If the stripping factor is not known or selected, determine the minimum air-to-water ratio by using Eq. 6.25. Obtain the design air-to-water ratio by multiplying this minimum air-to-water ratio with a value between 2 and 10. Go to Step 4.

Step 4: Determine the required air flow rate by multiplying the liquid flow rate with the air-to-water ratio determined from Step 2 or Step 3.

Example 6.7: Determine the Air-to-Water Ratio for an Air Stripper

A packed-column air stripper is designed to reduce the chloroform concentration in the extracted groundwater. Determine (a) the minimum air-to-water ratio, (b) the design air-to-water ratio, and (c) the design air flow rate. Use the following information in the calculations:

- Henry's constant for chloroform at 15°C, H = 2.12 atm/M
- stripping factor = 3
- extracted groundwater flow rate = 0.45 m³/min

SOLUTION:

(a) Use the formula in Table 2.3 to convert the Henry's constant to its dimensionless value:

$$H^* = \frac{H}{RT} = \frac{(2.12\,atmL\,/\,mol)}{(0.082\,atmL\,/\,(K\,mol))(273.15+15K)} = 0.0895$$

(b) Calculate the air-to-water ratio given by the stripping factor, using Eq. 6.26:

$$S = 3 = H^*\left(\frac{Q_a}{Q_w}\right) = (0.0895)\left(\frac{Q_a}{Q_w}\right)$$

$$\left(\frac{Q_a}{Q_w}\right) = 33.5$$

(c) Determine the required air flow rate by multiplying the liquid flow rate with the air-to-water ratio:

$$Q_a = Q_w\left(\frac{Q_a}{Q_w}\right) = \left(0.45\frac{m^3}{min}\right)(33.5) = 15\ m^3/min$$

Discussion: A stripping factor of 3 means the ratio of design to minimum Q_a/Q_w is also 3. Consequently, the design (Q_a/Q_w) can be obtained by multiplying the minimum Q_a/Q_w with the stripping factor.

Column Diameter. A key design parameter for an air stripper is the diameter of the column. The diameter depends mainly on the liquid flow rate. The higher the liquid flow rate is, the larger the column diameter would be. A typical liquid hydraulic loading rate, Q_L, for an air stripping column is 20 gpm/ft² (or 2.7 ft/min or 0.81 m/min) or less. This parameter is often used to determine the required cross-sectional area of the stripping column ($A_{stripping}$):

$$A_{stripping} = \frac{Q_w}{Q_L} \tag{6.27}$$

Packing Height. Another design parameter is the required height of the packing column (Z) for a specific removal efficiency. The taller the column, the larger the COC removal efficiency. The packing height can be determined using the transfer unit concept:

$$Z = (HTU) \times (NTU) \tag{6.28}$$

where HTU is the height of the transfer unit and NTU is the number of transfer units.

The HTU value depends on the hydraulic loading rate and the overall liquid-phase mass transfer coefficient, $K_L a$. K_L is the rate constant (m/sec) and "a" is the specific surface area (m²/m³). $K_L a$ has a unit of time^{-1}. The $K_L a$ value for a specific application can be best determined from pilot testing, and there are also empirical equations available to estimate the value of $K_L a$. Values of $K_L a$ in air stripping columns used in groundwater remediation range from 0.01 to 0.05 sec^{-1}. HTU has a unit of length and can be determined as:

$$HTU = \frac{Q_L}{K_L a} \tag{6.29}$$

where Q_L is the hydraulic loading rate in dimensions of length/time.

The NTU value can be determined by using the following formula:

$$NTU = \left(\frac{S}{S-1}\right) \ln\left[\frac{(C_{in} - G_{in}/H^*)}{(C_{out} - G_{in}/H^*)}\left(\frac{S-1}{S}\right) + \frac{1}{S}\right] \tag{6.30}$$

When $G_{in} = 0$, Eq. 6.30 simplifies to:

$$NTU = \left(\frac{S}{S-1}\right) \ln\left[\left(\frac{C_{in}}{C_{out}}\right)\left(\frac{S-1}{S}\right) + \frac{1}{S}\right] \tag{6.31}$$

where S is the stripping factor, H* is the dimensionless Henry's constant, C is the COC concentration in the water, and G is the COC concentration in air.

The following procedure can be used to size an air stripping column:

Step 1: Determine the required cross-sectional area of the air stripper by using Eq. 6.27. Then, determine the diameter of the column corresponding to this calculated area. Round up the diameter value.

Step 2: Use the newly-found diameter to calculate the design cross-sectional area and the hydraulic loading rate, Q_L. Use Eq. 6.29 to find the HTU value.

Step 3: Determine the stripping factor, if not known or specified, by using Eq. 6.25.

Step 4: Use Eq. 6.31 to find the NTU value.

Step 5: Use Eq. 6.28 to find the packing height, Z.

Example 6.8: Sizing an Air Stripper for Groundwater Remediation

A packed-column air stripper is designed to reduce chloroform concentration in the extracted groundwater. The concentration is to be reduced from 50 mg/L to

0.05 mg/L (50 ppb). Size the air stripper by determining the cross-sectional surface area, packing height, and air flow rate.

Use the following information in calculations:

- Henry's dimensionless constant for chloroform = 0.090
- stripping factor = 3
- extracted groundwater flow rate = 0.45 m³/min
- $K_La = 0.01$ s^{-1}
- hydraulic loading rate = 0.81 m/min
- chloroform concentration in the influent air = 0

SOLUTION:

(a) Use Eq. 6.27 to determine the required cross-sectional area:

$$A_{stripping} = \frac{Q_w}{Q_L} = \frac{0.45\dfrac{m^3}{min}}{0.81\ m/min} = 0.56\ m^2$$

The diameter of the air stripping column is

$$D = \left(\frac{4A}{\pi}\right)^{1/2} = \left(\frac{4(0.56)}{\pi}\right)^{1/2} = 0.84\ m,\ \text{which can be rounded up to 0.90 m}$$

(b) Use this newly-found diameter to find the hydraulic loading rate. The actual cross-sectional area of the column is now

$$A_{design} = \frac{\pi D^2}{4} = \frac{\pi(0.9)^2}{4} = 0.64\ m^2$$

The hydraulic loading rate to the column is

$$Q_L = \frac{Q_w}{A} = \frac{0.45}{0.64} = 0.71\ m/min$$

(c) Use Eq. 6.29 to determine the HTU value:

$$HTU = \frac{Q_L}{K_La} = \frac{0.71\ m/min}{0.01/s} \times \frac{1\ min}{60\ s} = 1.18\ m \approx 1.2\ m$$

(d) Use Eq. 6.31 to determine the NTU value:

$$NTU = \left(\frac{S}{S-1}\right)\ln\left[\left(\frac{C_{in}}{C_{out}}\right)\left(\frac{S-1}{S}\right) + \frac{1}{S}\right] = \left(\frac{3}{3-1}\right)\ln\left[\left(\frac{50}{0.05}\right)\left(\frac{3-1}{3}\right) + \frac{1}{3}\right] = 9.75$$

(e) Use Eq. 6.28 to determine the packing height:

$$Z = HTU \times NTU = 1.2 \times 9.75 = 11.7\ m$$

DISCUSSION:

1. The typical hydraulic loading rate (0.8 m/min or 20 gpm/ft^2) is much higher than that for the activated carbon adsorbers (0.2 m/min or 5 gpm/ft^2).
2. The required packing height of 11.7 m (38 ft) makes the total height of the air stripper very tall. This may not be acceptable in most project locations. If this is the case, one may consider having two shorter air strippers in series.

6.3.3 ADVANCED OXIDATION

Advanced oxidation processes (AOP) refer to a chemical oxidation process assisted by ultraviolet (UV) irradiation. This is also known as UV/oxidation. In AOP, the chemical oxidation process is the same as the *in situ* chemical oxidation (ISCO) process described in Section 5.5. An oxidizing agent, typically hydrogen peroxide, ozone, or a combination of these two, is injected into the system. The oxidizing agent is activated by the UV light to form hydroxyl radicals, which have a very strong oxidizing power. These radicals destroy the organic COCs in the impacted groundwater. The UV can also break the chemical bonds in the COC molecules by photolysis. The UV component consists of high-power lamps that emit UV radiation into the impacted groundwater.

AOP can be advantageous over carbon adsorption and air stripping because it directly destroys chemicals instead of simply transferring the COC from the water into a solid (carbon adsorption) or a gas (air stripping). The disadvantages of AOP are the high operating costs of UV lamps, the safety concerns of storing oxidizing agents on site, and other impurities in the water that can render the chemical reactions less efficient.

In a typical AOP, the oxidizing reagents are injected and mixed using metering pumps and in-line static mixers. The pumped groundwater then flows sequentially through one or more UV reactors. The reactors are often considered plug-flow, and the reactions follow first-order kinetics. Eq. 6.32 describes the relationship among the influent concentration, effluent concentration, hydraulic residence time, and reaction rate constant for first-order plug flow reactors (PFRs) and applied here for AOP reactors:

$$\frac{C_{out}}{C_{in}} = e^{-kt} = e^{-k(V/Q)} \tag{6.32}$$

where C is the COC concentration in the groundwater, V is the reactor volume, Q is the extracted groundwater flow rate, k is the reaction rate constant, and t is the hydraulic residence time. The reaction rate constant is a parameter determined by AOP equipment designers or determined in pilot studies.

Example 6.9: Designing the Reactors for an Advanced Oxidation Process

UV and ozone treatment is selected to remove TCE from an extracted groundwater stream (TCE concentration = 450 ppb). A pilot study was conducted with one *ex situ* reactor designed for a hydraulic residence time of 2 min. This system could reduce TCE concentration from 450 ppb to 25 ppb. However, the discharge limit for TCE is 5 ppb. Assuming the reactors behave as ideal plug-flow reactors and that the reaction is of first-order, how many of these reactors would you recommend to use? What would be the final TCE concentration with the number of recommended reactors?

SOLUTION:

(a) First determine the reaction rate constant, k, using Eq. 6.32:

$$\frac{C_{out}}{C_{in}} = e^{-kt}$$

$$\frac{25}{450} = e^{-2k}$$

$$\ln\left(\frac{25}{450}\right) = \ln(e^{-2k})$$

$$-2.89 = -2k$$

$$k = 1.44 \text{ min}^{-1}$$

(b) Then use Eq. 6.32 again to determine the required hydraulic residence time to reduce the TCE concentration to below the discharge limit:

$$\frac{5}{450} = e^{-1.44t}$$

$$t = 3.1 \text{ min}$$

Because each reactor has a residence time of 2 min, then we must recommend two reactors to achieve the required residence time of 3.1 min.

(c) Because two reactors will have a total residence time of 4 min, use Eq. 6.32 again with t = 4 min to determine the final effluent TCE concentration:

$$\frac{C_{out}}{450} = e^{-144(4)}$$

$$C_{out} = 1.4 \text{ ppb}$$

A C_{out} of 1.4 ppb is less than the discharge limit of 5 ppb, so two reactors provide sufficient removal of TCE.

DISCUSSION:

1. For PFRs, the final concentration from two identical reactors in series is the same as that from two identical reactors in parallel.
2. A pilot-scale test to determine the removal efficiency and the reaction rate constant is always recommended for AOPs.

6.3.4 CHEMICAL PRECIPITATION

Chemical precipitation is a common method to remove inorganic heavy metals from water. Chemical precipitation consists of passing extracted groundwater through a completely mixed flow reactor (CMFR) in which the water pH is increased to form hydroxides of heavy metals. To increase the pH, lime or caustic soda is added to the reactor. With the increased pH and formation of insoluble metal hydroxides, the metals precipitate out of the solution by settling to the bottom of the reactor. This is a precipitation reaction, and the solubility of metal hydroxides is sensitive to pH. The precipitation reaction can be expressed in the general form:

$$M^{n+} + nOH^- \leftrightarrow M(OH)_n \downarrow \qquad (6.33)$$

where M represents the heavy metal, OH^- is the hydroxide ion added to increase the pH, n is the valence of the metal, $M(OH)_n$ is the metal hydroxide, and the ($\downarrow$) represents the metal hydroxide in precipitate form.

The equilibrium equation can be written as:

$$K_{sp} = \left[M^{n+} \right]\left[OH^- \right]^n \qquad (6.34)$$

where K_{sp} is the equilibrium constant (also called the solubility product), $[M^{n+}]$ is the molar concentration of the heavy metal, and $[OH^-]$ is the molar concentration of hydroxide ions.

The K_{sp} values for $Cr(OH)_3$, $Fe(OH)_3$, $Mg(OH)_2$, and $Cu(OH)_2$ at 25°C are 6×10^{-31} M⁴, 6×10^{-36} M⁴, 9×10^{-12} M³, and 2×10^{-19} M³, respectively.

It is important to remember the definition of pH and pOH and the self-ionization constant of water, K_w:

$$pH = -\log\left[H^+ \right] \qquad (6.35)$$

$$pOH = -\log\left[OH^- \right] \qquad (6.36)$$

$$K_w = \left[H^+ \right]\left[OH^- \right] = 10^{-14} \qquad (6.37)$$

$$pH + pOH = 14 \qquad (6.38)$$

Example 6.10: Chemical Precipitation for Magnesium Removal

An extracted groundwater stream flows at 150 gpm and has a magnesium ion concentration of 100 mg/L. Sodium hydroxide is added to a CMFR to a pH of 11 to precipitate and remove magnesium ion from the water.

- The temperature of the reactor is kept at 25°C
- The solubility product of $Mg(OH)_2$ is 9×10^{-12} M³ at 25°C

- MW of Mg = 24.3 g/mol
- MW of Mg(OH)$_2$ = 58.3 g/mol

Estimate the:

(a) Mg^{2+} concentration in the treated effluent (mg/L)
(b) rate of Mg(OH)$_2$ produced (lb/day)
(c) rate of sludge produced (lb sludge/day) if the solid content of the sludge is 10% by weight.

SOLUTION:

(a) First, write the precipitation reaction:

$$Mg^{2+} + 2OH^- \leftrightarrow Mg(OH)_2 \downarrow$$

Find the concentration of OH$^-$ at pH = 11. First, find pOH:

$$11 + pOH = 14$$

$$pOH = 3$$

Then find [OH$^-$]:

$$pOH = -\log\left[OH^-\right]$$

$$3 = -\log\left[OH^-\right]$$

$$\left[OH^-\right] = 10^{-3} \text{ mol/L}$$

Use the solubility product equation to determine the magnesium concentration at pH 11:

$$K_{sp} = \left[Mg^{2+}\right]\left[OH^-\right]^2$$

$$9 \times 10^{-12} = \left[Mg^{2+}\right]\left(10^{-3}\right)^2$$

$$\left[Mg^{2+}\right] = 9 \times 10^{-6} \text{ M}$$

Convert the molar concentration to mg/L. This is the concentration of Mg^{2+} in the treated effluent.

$$C_{Mg^{2+}} = \left[Mg^{2+}\right] \times MW = 9 \times 10^{-6} \frac{mol}{L} \times 24.3 \frac{g}{mol} \times \frac{1000 \text{ mg}}{g} = 0.22 \frac{mg}{L}$$

(b) As shown in the chemical equation in part (a), one mole of $Mg(OH)_2$ is formed for each mole of Mg^{2+}. The rate of $Mg(OH)_2$ produced can be found as:

$$\text{Rate of } Mg(OH)_2 \text{ produced} = \left(\text{Rate of } Mg^{2+} \text{removed}\right)\left(\text{Molar ratio of } Mg(OH)_2 \text{ to } Mg^{2+}\right)$$

$$= \left\{\left[Mg^{2+}\right]_{in} - \left[Mg^{2+}\right]_{out}\right\} \times (Q) \times (58.3 / 24.3)$$

$$= \left[(100 - 0.22) \text{ mg}/L\right] \times \left[(150 \text{ gpm})(3.785 \text{ L}/\text{gal})\right] \times (58.3 / 24.3)$$

$$= 136{,}000 \text{ mg/min} = 136 \text{ g/min} = 431 \text{ lb/day}.$$

(c) Since the solids are settled to 10% by weight, the rate of sludge production can be found as:

$$\text{Rate of sludge produced} = \text{Rate of } Mg(OH)_2 \text{ produced} \div 10\%$$

$$= 431 \text{ lb/day} \div 10\% = 4{,}310 \text{ lb/day}.$$

6.4 *IN SITU* GROUNDWATER REMEDIATION

6.4.1 AIR SPARGING

Air sparging is an in-situ remediation technology that involves the injection of air into the saturated zone. The injected air travels through the aquifer, moves upward through the capillary fringe and the vadose zone, and is then collected by the vadose-zone soil vapor extraction network (Figure 6.13). The injected air: (i) volatilizes the

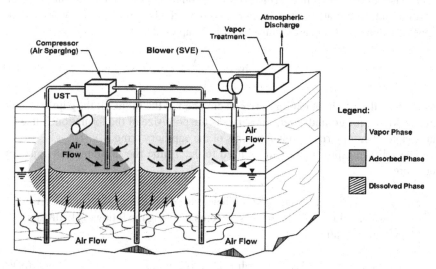

FIGURE 6.13 An air sparging system with three wells with injection points deep into the saturated zone, and a soil vapor extraction system with three extraction wells in the vadose zone. Note the mounding of the groundwater table around the air sparging wells, caused by the upward movement of air. (Source: U.S. EPA 2017a.)

dissolved VOCs in the groundwater, (ii) supplies oxygen to the aquifer for biore-mediation, (iii) volatilizes the VOCs in the capillary zone as it moves upward, and (iv) volatilizes the VOCs in the vadose zone. Because the air moves upward into the vadose zone, air sparging can only be used in unconfined aquifers, rather than confined aquifers. Air sparging is most effective in soils with intrinsic permeability larger than 0.1 darcy (10^{-9} cm^2) and COCs with Henry's constant greater than 100 atm and vapor pressures higher than 0.5 mm-Hg (U.S. EPA 2017a).

How many air sparging wells to install, and how far apart they should be, depends on pilot tests in the field. The air flow ranges from 3 to 25 standard cubic feet per minute (scfm) (0.08 to 0.7 m^3/min at standard temperature and pressure) per injection well (U.S. EPA 2017a).

Injection Pressure of Air Sparging

To design an air sparging system, a key parameter is the air injection pressure. The applied air injection pressure should overcome at least (i) the hydrostatic pressure corresponding to the water column height above the injection point and (ii) the "air entry pressure," which is equivalent to the capillary pressure necessary to induce air into the saturated media.

$$P_{injection} = P_{hydrostatic} + P_{capillary} \qquad (6.39)$$

Reported values of injection pressures range from 1 to 8 psig (7 to 55 kPa) (Johnson et al. 1993). Note that these pressure values are in terms of gauge pressure and not absolute pressure.

The height of the capillary fringe at a site strongly depends on its subsurface geology. For pure water at 20°C in a clean glass tube, the height of capillary rise can be approximated by the equation introduced in Chapter 4:

$$h_c = \frac{0.153}{r} \qquad (4.3)$$

where h_c is the height of capillary rise in cm and r is the radius of the capillary tube in cm, which, for groundwater, can serve as the pore size of the formation. Table 4.1 in Chapter 4 summarizes typical heights of the capillary fringe.

The following procedure can be used to determine the minimum air injection pressure:

Step 1: Determine the water column height above the injection point. Convert the water column height to pressure units by using the following hydrostatic pressure equation:

$$P_{hydrostatic} = \gamma h_{hydrostatic} = \rho g h_{hydrostatic} \qquad (6.40)$$

where γ is the specific weight of water, h is the height of the water column above the injection point, ρ is the mass density of water, and g is the acceleration due to gravity.

Step 2: Use Table 4.1 to estimate the pore radius of the aquifer media and then use Eq. 4.3 to determine the height of the capillary rise (or obtain the capillary height from Table 4.1 directly). Convert the capillary height to the capillary pressure by using the following formula:

$$P_{capillary} = \gamma h_{capillary} = \rho g h_{capillary} \tag{6.41}$$

Typical values for the parameters in the above equation are:
At $T = 60°F$: $\gamma = 62.36$ lbf/ft³; $\rho = 62.36$ lbm/ft³
At $T = 20°C$: $\gamma = 9.789$ kN/m³; $\rho = 998.2$ kg/m³
$g = 32.2$ ft/s² $= 9.81$ m/s²

Step 3: The minimum air injection pressure is the sum of the above two pressure components.

The following two examples show these calculations, the first one in the U.S. customary system of units, and the second one in SI units.

Example 6.11: Determine the Required Injection Pressure of Air Sparging, U.S. Customary Units

Three air sparging wells are to be installed into a groundwater plume. The water is at a temperature of 60°F. The injection air flow rate into each well is 5 ft³/min. The height of the water column above the air injection point is 10 ft. Determine the minimum air injection pressure, in psig units, required if (a) the aquifer matrix consists mainly of coarse sand and (b) the aquifer formation is clayey. Compare the two pressures.

SOLUTION:

(a) Use Eq. 6.40 to convert the water column height to pressure units as:

$$P_{hydrostatic} = \gamma h_{hydrostatic} = 62.36 \frac{lbf}{ft^3} \times 10\,ft = 624 \frac{lbf}{ft^2}$$

or

$$P_{hydrostatic} = \rho g h_{hydrostatic} = 62.36 \frac{lbm}{ft^3} \left(32.2 \frac{ft}{s^2}\right)(10\,ft)\left(\frac{1lbf}{32.2lbm\frac{ft}{s^2}}\right) = 624 \frac{lbf}{ft^2}$$

Note in the equation above the conversion from lbm to lbf. Converting the answer above to psi:

$$624 \frac{lbf}{ft^2} \times \left(\frac{1ft}{12in}\right)^2 = 4.33 \frac{lbf}{in^2} = 4.33 psi$$

From Table 4.1, the pore radius of fine sand media is 0.05 cm. Use Eq. 4.3 to determine the height of the capillary rise:

$$h_c = \frac{0.153}{r} = \frac{0.153}{0.05} = 3.06 \text{ cm} = 0.10 \text{ ft}$$

Calculate the capillary pressure from the capillary rise:

$$P_{capillary} = \gamma h_{capillary} = 62.36 \frac{lbf}{ft^3} \times 0.10 \text{ ft} = 6.26 \frac{lbf}{ft^2} = 0.0433 \text{ psi}$$

Use Eq. 6.39 to determine the minimum air injection pressure:

$$P_{injection} = P_{hydrostatic} + P_{capillary} = 4.33 + 0.0433 = 4.37 \text{ psig}$$

(b) If the aquifer formation is clayey, then the pore radius is 0.0005 cm from Table 4.1. Use Eq. 4.3 to determine the height of the capillary rise:

$$h_c = \frac{0.153}{r} = \frac{0.153}{0.0005} = 306 \text{ cm} = 10 \text{ ft}$$

Calculate the capillary pressure from the capillary rise:

$$P_{capillary} = \gamma h_{capillary} = 62.36 \frac{lbf}{ft^3} \times 10 \text{ ft} = 623.6 \frac{lbf}{ft^2} = 4.33 \text{ psi}$$

Use Eq. 6.39 to determine the minimum air injection pressure:

$$P_{injection} = P_{hydrostatic} + P_{capillary} = 4.33 + 4.33 = 8.66 \text{ psig}$$

Discussion:

1. For sandy aquifers, the air injection pressure is close to the hydrostatic pressure, and the capillary pressure, or "air entry pressure," is negligible. However, for clayey aquifers, the air injection pressure is highly influenced by both the hydrostatic pressure and the air entry pressure because of the large capillary rise.
2. The actual air injection pressure should be larger than the minimum air injection pressure calculated above to cover the system pressure loss such as head loss in the pipeline, fittings, and injection head.
3. The calculated injection pressures are in the ballpark of the reported field values, 1–8 psig.
4. 1 atm = 14.696 psi = 33.98 ft-H_2O.

Example 6.12: Determine the Required Injection Pressure of Air Sparging, SI Units

Three air sparging wells are to be installed into a groundwater plume. The water is at a temperature of 20°C. The injection air flow rate into each well is 0.14 m³/min.

The height of the water column above the air injection point is 3 m. Determine the minimum air injection pressure, in kPa units, required if (a) the aquifer matrix consists mainly of coarse sand and (b) the aquifer formation is clayey. Compare the two pressures.

SOLUTION:

(a) Use Eq. 6.40 to convert the water column height to pressure units as:

$$P_{hydrostatic} = \gamma h_{hydrostatic} = 9.789 \frac{kN}{m^3} \times 3\,m = 29.4 \frac{kN}{m^2} = 29.4\,kPa$$

or

$$P_{hydrostatic} = \rho g h_{hydrostatic} = 998.2 \frac{kg}{m^3} \left(9.81 \frac{m}{s^2}\right)(3\,m) = 29377 \frac{N}{m^2} = 29.4\,kPa$$

From Table 4.1, the pore radius of fine sand media is 0.05 cm. Use Eq. 4.3 to determine the height of the capillary rise:

$$h_c = \frac{0.153}{r} = \frac{0.153}{0.05} = 3.06\,cm = 0.0306\,m$$

Calculate the capillary pressure from the capillary rise:

$$P_{capillary} = \gamma h_{capillary} = 9.789 \frac{kN}{m^3} \times 0.0306\,m = 0.2995 = 0.30\,kPa$$

Use Eq. 6.39 to determine the minimum air injection pressure:

$$P_{injection} = P_{hydrostatic} + P_{capillary} = 29.4 + 0.30 = 29.7\,kPa$$

(b) If the aquifer formation is clayey, then the pore radius is 0.0005 cm from Table 4.1. Use Eq. 4.3 to determine the height of the capillary rise:

$$h_c = \frac{0.153}{r} = \frac{0.153}{0.0005} = 306\,cm = 3.06\,m$$

Calculate the capillary pressure from the capillary rise:

$$P_{capillary} = \gamma h_{capillary} = 9.789 \frac{kN}{m^3} \times 3.06\,m = 29.95\,kPa$$

Use Eq. 6.39 to determine the minimum air injection pressure:

$$P_{injection} = P_{hydrostatic} + P_{capillary} = 29.4 + 29.95 = 59.3\,kPa$$

DISCUSSION:

1. For sandy aquifers, the air injection pressure is close to the hydrostatic pressure, and the capillary pressure, or "air entry pressure," is negligible. However, for clayey aquifers, the air injection pressure is highly

influenced by both the hydrostatic pressure and the air entry pressure because of the large capillary rise.

2. The actual air injection pressure should be larger than the minimum air injection pressure calculated above to cover the system pressure loss, such as head loss in the pipeline, fittings, and injection head.

3. Converting the answers from Example 6.11 to Example 6.12 shows that the values are similar and differ only by the rounding of numbers. Comparing both examples is a good exercise in understanding the U.S. customary system of units and SI units.

4. 1 atm = 101.3 kPa = 10.33 m-H_2O.

6.4.2 IN SITU BIOREMEDIATION

One of the most common types of *in situ* remediation is the bioremediation of organic COCs in aquifers by enhancing the metabolism of indigenous microorganisms present in the subsurface. Most *in situ* bioremediation is practiced in the aerobic mode. As introduced in Chapter 5, bioremediation is often practiced as the biostimulation of the metabolism of microbes. The biostimulation is carried out by the addition of oxygen or inorganic nutrients into the groundwater plume.

Addition of Oxygen to Enhance Biodegradation

Groundwater naturally contains low levels of dissolved oxygen (DO). Even if it is fully saturated with air, the saturated dissolved oxygen (DO_{sat}) concentration in groundwater would only be approximately 9 mg/L at 20°C. The biodegradation of organic COCs in the plume requires much more than that amount of DO.

The addition of oxygen to the groundwater can be done by air sparging (see Section 6.4.1) or pure oxygen sparging. The oxygen in the injected air can raise the DO to its saturation level of 8–10 mg/L. With pure oxygen injection, DO concentrations of up to 40 or 50 mg/L can be achieved.

Oxygen Addition from Air Sparging

Air sparging was discussed earlier as a mass transfer process, where air injected into the aquifer volatilizes COCs, transporting those COCs out of the water and into the air stream that moves from the aquifer to the vadose zone. But air sparging also serves the function of adding oxygen into the aquifer to promote bioremediation. This process is called biosparging. Consequently, supplying oxygen to the plume to support aerobic biodegradation is one of the main functions of air sparging.

The oxygen transfer efficiency (OTE) is often used to evaluate the efficacy of aeration and is defined as:

$$\text{Oxygen Transfer Efficiency}\left(\text{OTE}\right) = \frac{\text{Rate of Oxygen Dissolution, } R_d}{\text{Rate of Oxygen Applied, } R_a} \quad (6.42)$$

The OTE should depend on factors such as injection pressure, the depth of the injection point in the aquifer, and characteristics of the geological formation, to name a few. This is not a highly efficient process, around 10% efficient.

The rate of oxygen applied (R_a) can be calculated using the following steps.

Step 1. The oxygen concentration in the ambient air is approximately 21% by volume, which is equal to 210,000 ppmV. Convert this concentration to a mass concentration, G, using Eqs. 2.5 or 2.6.

Step 2. Calculate the rate of application, which is equal to a mass flow rate: $R_a = G \times Q_{air}$.

Example 6.13: Determine the Rate of Oxygen Addition by Air Sparging

A biosparging well was installed into the plume of an aquifer impacted by hydrocarbons. The injection air flow rate into the well is 0.14 m³/min. Assuming the oxygen transfer efficiency (OTE) is 10%, determine the rate of oxygen addition to the aquifer through the sparging well. What would be the equivalent injection rate of water with a dissolved oxygen (DO) concentration of 9 mg/L?

SOLUTION:

(a) The oxygen concentration in the ambient air at 1 atm and T=20°C is approximately 21%, or 210,000 ppmV, which can be converted to a mass concentration using Eq. 2.5.

$$G\left(in \frac{mg}{m^3}\right) = G(in\ ppmV)\left(\frac{MW}{MV(inL)}\right)$$

The molar volume (MV) of ambient air is determined using the Ideal Gas Law, Eq. 2.7:

$$V = \frac{nRT}{P} = \frac{(1)\left(0.082\frac{L \cdot atm}{mol \cdot K}\right)(273.15 + 20\,K)}{1atm} = 24.04\,L$$

Inserting this information back into Eq. 2.5:

$$G\left(in \frac{mg}{m^3}\right) = 210,000(in\ ppmV)\left(\frac{32\frac{g}{mol}}{24.04\frac{L}{mol}}\right) = 279,534\ mg/m^3$$

The rate of oxygen injected in each well is

$$R_a = G \times Q_{air} = 279,534\frac{mg}{m^3} \times 0.14\frac{m^3}{min} = 3.91 \times 10^4\frac{mg}{min} = 39.1\frac{g}{min} = 56\frac{kg}{day}$$

The rate of oxygen dissolved into the plume (R_d) through air injection in each well (using Eq. 6.42) = (56 kg/day)(10%) = 5.6 kg/day

(b) The DO concentration of the air-saturated reinjection water is approximately 9 mg/L at 20°C. The rate of oxygen transferred from water is also a mass flow rate:

$$\dot{M} = R_d = C \times Q$$

If the required rate of oxygen supply by water is $\dot{M} = R_d = 5.6$ kg/day, and the DO concentration in the water is $C = 9$ mg/L, then

$$Q = \frac{\dot{M}}{C} = \frac{5.7 \frac{kg}{day}}{9 \frac{mg}{L}} \times \frac{10^6 \, mg}{kg} = 6.22 \times 10^5 \frac{L}{day} = 622 \frac{m^3}{day} = 0.43 \frac{m^3}{min}$$

DISCUSSION:

1. The oxygen transfer efficiency of 10% means that only 10% of the total oxygen sparged into the aquifer is dissolved into the aquifer. But, 90% of the oxygen injected can serve as the oxygen source for bioremediation in the vadose zone.
2. Despite the relatively low oxygen transfer efficiency in this example, air sparging still adds a significant amount of oxygen to the aquifer. With regard to oxygen addition, an air injection rate of 0.14 m³/min with an oxygen transfer efficiency of 10% is equivalent to the injection of air-saturated water at 0.43 m³/min.

The DO level in the water can also be raised by the addition of chemicals, such as hydrogen peroxide (H_2O_2) and ozone (O_3). In this case, groundwater is extracted, mixed with H_2O_2 or O_3, and reinjected into the aquifer. Each mole of hydrogen peroxide in water can dissociate into half a mole of oxygen and one mole of water, while one mole of ozone in water can dissociate into one and a half moles of oxygen as:

$$H_2O_2 \rightarrow H_2O + 0.5 \, O_2 \tag{6.43}$$

$$O_3 \rightarrow 1.5 \, O_2 \tag{6.44}$$

Ozone is ten times more soluble in water than pure oxygen and is a strong oxidant, along with hydrogen peroxide, as seen in Section 6.3.3. In addition to providing oxygen for biodegradation, ozone and hydrogen peroxide can also generate radicals to oxidize COCs and other inorganic and organic compounds present in the aquifer. At higher concentrations, though, they may become toxic to indigenous aerobic microorganisms and suppress their biological activities.

Other *in situ* biodegradation approaches rely on oxygen releasing compounds (ORCs). Common ORCs include calcium and magnesium peroxides that are introduced to the saturated zone in solid powder or a slurry through well. These peroxides release the oxygen to the aquifer when hydrated by groundwater passing through the wells. Magnesium peroxide has been more commonly used than calcium peroxide due to its lower solubility and prolonged release of oxygen. Oxygen amounting to ~10% of the mass of magnesium peroxide placed in the saturated zone is released to the aquifer over the active period (U.S. EPA 2017b).

Example 6.14: Determine the Effectiveness of Hydrogen Peroxide Addition as an Oxygen Source for Bioremediation

For bioremediation, a typical maximum concentration of hydrogen peroxide in the injected groundwater is 1,000 mg/L. Determine the mass concentration of oxygen that 1,000 mg/L of hydrogen peroxide can provide. Compare this concentration with the typical dissolved oxygen concentration found in groundwater.

SOLUTION:

(a) From Eq. 6.43, one mole of hydrogen peroxide can yield one-half mole of oxygen:

$$H_2O_2 \rightarrow H_2O + 0.5\ O_2$$

Molecular weight of hydrogen peroxide $(H_2O_2) = (1 \times 2) + (16 \times 2) = 34$ g/mol
Molecular weight of oxygen $(O_2) = 16 \times 2 = 32$ g/mol

(b) Molar concentration of 1,000 mg/L hydrogen peroxide

$$C_{H_2O_2} = \frac{1000\frac{mg}{L}}{34\ \text{g/mol}} \times \frac{1\,g}{1000\ mg} = 0.0294\frac{mol}{L} = 0.0294\ M$$

Molar concentration of oxygen (assume 100% dissociation of hydrogen peroxide), according to Eq. 6.43:

$$C_{O_2} = 0.0294\frac{mol}{L} \times \frac{0.5\ mol\ O_2}{1\ mol\ H_2O_2} = 0.0147\frac{mol}{L} = 0.0147\ M$$

Mass concentration of oxygen in water from hydrogen peroxide addition

$$C_{O_2,mass} = 0.0147\frac{mol}{L} \times 32\frac{g}{mol} \times \frac{1000\ mg}{1\,g} = 470\frac{mg}{L}$$

The DO_{sat} in groundwater is typically 9 mg/L. Therefore adding H_2O_2 contributes 470/9 = 52 times more oxygen for bioremediation. However, note that a typical DO is much lower than the DO_{sat}.

Addition of Nutrients to Enhance Biodegradation

In the subsurface, nutrients that enhance microbial activity usually already exist. However, with the presence of organic COCs, additional nutrients are often needed to support bioremediation. The nutrients to enhance microbial growth are assessed primarily on the nitrogen and phosphorus requirements of the microorganisms. The suggested C:N:P molar ratio is 120:10:1, as shown in Table 5.1 in Chapter 5. The nutrients are typically added at concentrations ranging from 0.005 to 0.02% by weight (U.S. EPA 1991). Much of the discussion for *in situ* groundwater bioremediation is similar to *in situ* vadose zone remediation, as described in Section 5.4. The difference is that now we are adding nutrients to stimulate the biodegradation of COCs in the soil matrix that is saturated with water.

For procedures to calculate the nutrient requirements for groundwater bioremediation, follow the steps below:

Step 1: Determine the moles of COC in a unit volume of aquifer.

Step 2: Determine the moles of C in the COC in the same unit volume of aquifer.

Step 3: Use the given C:N:P molar ratio to determine the moles of N and P needed.

Step 4: Given the moles of N and P, calculate the mass of N and P needed.

Step 5: Given the mass of N and P required, calculate the mass of the N and P compounds needed.

Example 6.15: Determine the Nutrient Requirement for *In Situ* Groundwater Bioremediation

A groundwater aquifer is impacted by gasoline. The average dissolved gasoline concentration of the groundwater samples is 20 mg/L. At this site, on average, 1 m³ of aquifer contains 225 g of gasoline. The porosity is 0.35. Assume gasoline has the same formula as heptane, C_7H_{16}. *In situ* bioremediation is being considered for aquifer restoration, using ammonium sulfate $(NH_4)_2SO_4$ and trisodium phosphate $Na_3PO_4 \cdot 12H_2O$ as the added N and P compounds.

Assuming no nutrients are available in the groundwater for bioremediation and that the optimal molar C:N:P ratio has been determined to be 100:10:1, determine the mass of nutrients needed to support the biodegradation of gasoline, on a basis of 1 m³ of aquifer. If the plume is to be flushed with 100 pore volumes of oxygen- and nutrient-enriched water, what would be the required nutrient concentration in the water for reinjection?

SOLUTION:

Basis: 1 m³ of aquifer

(a) Determine the moles of gasoline in 1 m³ of aquifer:
   MW of gasoline $(C_7H_{16}) = 7 \times 12 + 1 \times 16 = 100$ g/mol

$$Moles_{gasoline} = \frac{225 \text{ g/m}^3}{100 \text{ g/mol}} = 2.25 \text{mol/m}^3$$

(b) Determine the number of moles of C:
   Since there are 7 carbon atoms in each gasoline molecule, as indicated by its formula C_7H_{16}, then:

$$Moles_C = \left(2.25\frac{\text{mol}}{\text{m}^3}\right)(7) = 15.8\frac{\text{mol}}{\text{m}^3}$$

(c) Determine the mass of N and mass of $(NH_4)_2SO_4$ needed (using the C:N:P ratio of 100:10:1):

$$\text{Moles of N needed} = (10/100)(15.8) = 1.58\,\text{mol/m}^3$$

Mass of nitrogen needed $= 1.58\,\text{mol/m}^3 \times 14\,\text{g/mol} = 22.1\,\text{g/m}^3$ of aquifer

According to the formula $(NH_4)_2SO_4$, 2 moles of N are present for each mole of $(NH_4)_2SO_4$. So the number of moles of $(NH_4)_2SO_4$ needed are $= 1.58 \div 2 = 0.79\,\text{mol/m}^3$

Mass of $(NH_4)_2SO_4$ needed $= (0.79\,\text{mol/m}^3)\left[(14+4)(2)+32+(16)(4)\,\text{g/mol}\right]$

$$= 104\,\text{g/m}^3 \text{ of aquifer.}$$

(d) Determine the mass of P and mass of $Na_3PO_4 \cdot 12H_2O$ needed (using the C:N:P ratio of 100:10:1):

$$\text{Moles of P needed} = (1/100)(15.8) = 0.158\,\text{mol/m}^3$$

Mass of phosphorus needed $= 0.158\,\text{mol/m}^3 \times 31\,\text{g/mol} = 4.9\,\text{g/m}^3$ of aquifer

According to the formula $Na_3PO_4 \cdot 12H_2O$, 1 mole of P is present for each mole of $Na_3PO_4 \cdot 12H_2O$. So the number of moles of $Na_3PO_4 \cdot 12H_2O$ needed is $= 0.158\,\text{mol/m}^3$

Mass of $Na_3PO_4 \cdot 12H_2O$ needed

$$= (0.158\,\text{mol/m}^3)\left[(23)(3)+31+(16)(4)+(12)(18)\,\text{g/mol}\right]$$

$$= 60\,\text{g/m}^3 \text{ of aquifer}$$

(e) The total nutrient requirement is 104 g of $(NH_4)_2SO_4$ and 60 g of $Na_3PO_4 \cdot 12H_2O$ per m^3 of aquifer.

(f) Determine the volume of water in 100 pore volumes

$$\text{Void space of the aquifer} = V \times \phi$$

$$= (1m^3)(35\%) = 0.35\,m^3 = 350\,L$$

Total volume of water that is equivalent to 100 pore volumes $= (100)(350)$

$$= 35,000\,L$$

(g) Determine the nutrient concentration for 100 pore volumes:
The minimum required concentration for $(NH_4)_2SO_4$ = 104 g ÷ 35,000 L
= 0.00297 g/L = 2.97 mg/L ≈ 0.0003% by weight
The minimum required concentration for $Na_3PO_4 \cdot 12H_2O$ = 60 g ÷ 35,000 L = 0.00171 g/L = 1.71 mg/L ≈ 0.0002% by weight

Discussion: The concentration, totaling about 0.0005% by weight is the theoretical amount. In real applications, one may want to add more to compensate for the loss due to adsorption to the aquifer material before reaching the plume. This makes the nutrient concentration fall in the typical range of 0.005–0.02% by weight.

6.4.3 *In Situ* Chemical Oxidation

While advanced oxidation process (AOP) is an *ex situ* groundwater treatment technique, chemical oxidation can also be practiced *in situ* (*in situ* chemical oxidation, ISCO). In groundwater ISCO, the chemical oxidation process is the same as the vadose zone process described in Section 5.5. An oxidizing agent, typically hydrogen peroxide, ozone, or a combination of these two, is injected into the groundwater, as shown in Figure 6.14. The resulting radicals destroy the organic COCs in the impacted groundwater.

6.5 MONITORED NATURAL ATTENUATION

Natural attenuation is nature's remediation process to decompose and reduce the spread of COCs without active remediation. Natural attenuation does not involve the use of engineered products, such as pumps, air compressors, pipes, valves, gauges, and electronic sensors. Therefore it can also be known as passive remediation or intrinsic remediation. Monitored natural attenuation (MNA) is not a "do nothing" activity – it importantly includes the term "monitored" because action is necessary

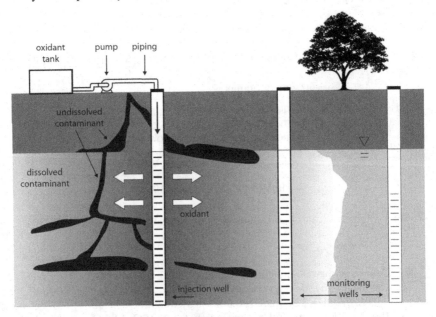

FIGURE 6.14 An illustration of the application of *in situ* chemical oxidation (ISCO), where an oxidant injected into a well spreads into the aquifer.

to collect soil and groundwater samples and analyze the data. These actions verify if natural attenuation is occurring as expected in order to protect human life and the environment. MNA is a valid remediation strategy when the source of contamination has been removed or controlled and if the contamination does not threaten a receptor (public supply wells, surface water bodies, basements) in the time necessary to achieve cleanup standards. MNA works for both vadose zone soil and groundwater, and we will focus here on groundwater.

6.5.1 Processes in Natural Attenuation

The following natural processes can occur underground to promote natural attenuation (U.S. EPA 2012c, 2017c):

Dispersion and Dilution

These processes spread and decrease the concentrations of COCs. They happen as the COCs move with groundwater and also as clean up-gradient groundwater mixes with the contaminated plume. This is a nondestructive technique (does not destroy the COCs into smaller compounds) but can render their concentrations below cleanup standards, albeit occupying a larger area.

Adsorption

Adsorption is the process of COCs sticking or adhering to soil particles. Adsorption mechanisms are described by the adsorption and partitioning discussion in Chapter 2. Adsorption is also nondestructive but can render COCs relatively immobile to not reach sensitive receptors.

Volatilization

This process is nondestructive and applies to volatile COCs when they leave the aquifer by changing phases from liquids to gases within the soil grains above the water table. These gases can also move up into the atmosphere, which is acceptable if their diluted concentrations are below air quality standards.

Biodegradation

Biodegradation is a natural bioremediation process that can occur without biostimulation. Indigenous microbes in the soil metabolize COCs and biodegrade them into water and less harmful compounds. Biodegradation can be aerobic or anaerobic and is carried out via oxidation–reduction reactions. Many times, biodegradation that starts as an aerobic reaction will deplete the soil and groundwater of oxygen, which is the electron acceptor. Terminal electron acceptors (TEAs) are the reduced compounds (that receive or gain electrons) that oxidize the COC (that is, the oxidized compound that donates or loses electrons). Where oxygen is depleted, anaerobic biodegradation can still happen if other electron acceptors are present, for example, nitrate, sulfate, ferric iron, manganese, and carbon dioxide (see Figure 6.15). Biodegradation is a destructive technique because it reduces the total mass of the COC.

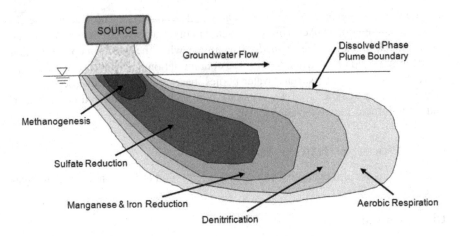

FIGURE 6.15 Conceptualization of terminal electron acceptor (TEA) zones in a contaminated groundwater plume. (Source: NJDEP 2012.)

Chemical Reactions

Chemical reactions are also destructive techniques, where COCs reacting with naturally occurring chemicals are rendered less toxic.

6.5.2 Designing a Monitoring Plan for Natural Attenuation

Initially, proper site characterization must be performed according to concepts introduced in Chapter 4. The site characterization should conclude with a site conceptual model that includes: groundwater flow direction; hydraulic gradient; hydraulic conductivity; plume boundaries; soil types; soil bulk densities; soil organic carbon fraction; and COC mass in the free product, vadose zone, and saturated zone.

After the initial site characterization, additional monitoring wells must be installed for MNA. Monitoring wells should be placed at the following locations (Wisconsin DNR 2014) and as illustrated in Figures 6.16 and 6.17:

- Up-gradient of the plume to determine the water quality entering the source area.
- At the source area to determine whether the source is increasing, stable, or decaying.
- At the center flowline of the plume to determine the general movement of COCs in the main direction of flow.
- Down-gradient of the plume to detect COCs approaching a sensitive receptor.
- Side-gradient of the plume to determine the width of the plume, whether it is expanding, and whether the groundwater flow direction is fluctuating.

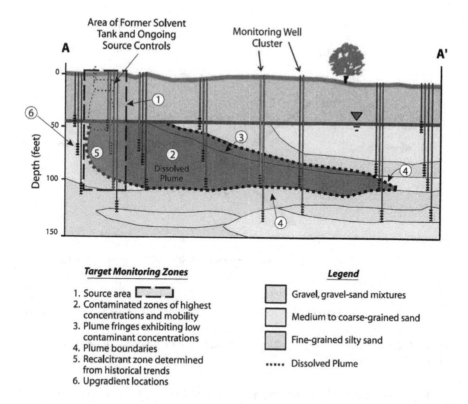

Target Monitoring Zones

1. Source area
2. Contaminated zones of highest concentrations and mobility
3. Plume fringes exhibiting low contaminant concentrations
4. Plume boundaries
5. Recalcitrant zone determined from historical trends
6. Upgradient locations

Legend

☐ Gravel, gravel-sand mixtures

☐ Medium to coarse-grained sand

☐ Fine-grained silty sand

····· Dissolved Plume

FIGURE 6.16 An example of a plume cross-section and locations of monitoring wells for MNA. (Source: Pope et al. 2004.)

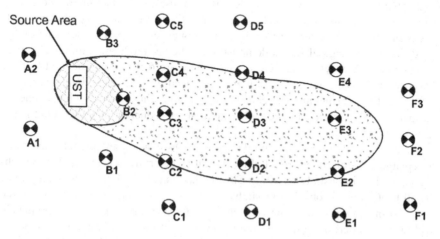

FIGURE 6.17 An example of a plume plan view and locations of monitoring wells for MNA. (Source: U.S. EPA 2017b.)

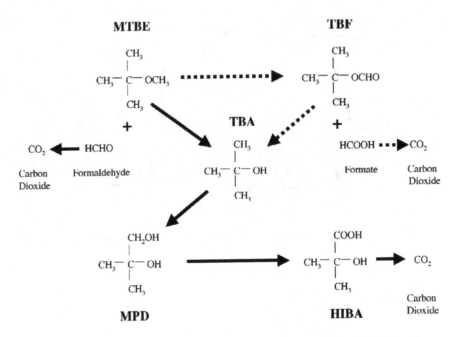

FIGURE 6.18 Aerobic biodegradation pathway and by-products of MTBE. (Source: ITRC 2005.)

With periodic (monthly or quarterly) monitoring of groundwater elevations and water quality at these monitoring wells, the hydrogeologic parameters introduced in Chapter 4 can be measured.

In addition to determining locations of monitoring wells to determine hydrogeologic characteristics, water quality parameters must also be measured. Measuring COC concentrations in each well helps determine the extent of the plume and its decay rate in groundwater. Further, by analyzing water samples for degradation by-products of the original COCs demonstrates whether biodegradation or chemical reactions are destroying the COCs (e.g., see Figure 6.18 showing methyl tertiary butyl ether, MTBE, as the COC, which aerobically biodegrades into several by-products). Even further analysis can be done by collecting samples of biogeochemical parameters, such as organic carbon, oxidation–reduction potential (ORP), and TEAs.

An estimate of the time it will take for the COC concentration to decrease to an acceptable concentration through dilution can be determined using the batch flushing method (Wisconsin DNR 2014). This method makes the following assumptions: (1) all COCs are dissolved in groundwater; (2) no COCs are present in the source area; (3) incoming groundwater is free of COCs and has time to mix completely within the aquifer; and (4) COC reductions happen due to dilution alone (and no other natural attenuation processes). The following steps describe the calculations for this assessment.

Step 1: Calculate the number of pore volumes to flush the COC, given by

$$PV = -R \ln\left(\frac{C_f}{C_i}\right) \tag{6.45}$$

where PV is the pore volume (unitless), R is the retardation factor (unitless), C_f is the desired cleanup concentration, and C_i is the initial concentration.

Step 2: Calculate the time for incoming groundwater to flow through the plume, given by

$$\tau = \frac{L}{v} \tag{6.46}$$

where τ is the groundwater flow time, L is the length of the COC plume, and v is the seepage velocity.

Step 3: Calculate the time for the number of calculated pore volumes to flow through the plume, given by

$$T = (PV) \times \tau \tag{6.47}$$

Where T is the time it will take to reduce the COC concentration from C_i to C_f due to flushing/dilution alone.

Step 4: If necessary, the first-order decay rate, k, (time^{-1}) can be calculated as $k = \frac{1}{R\tau}$. The first-order decay reaction is

$$C_f = C_i e^{-kT} \tag{6.48}$$

where T is the previously calculated total time of cleanup.

Example 6.16: Determine the Time to Reduce a Groundwater COC Concentration through Flushing and Dilution.

A benzene plume 58 m in length at an average concentration of 15,000 ppb is to be assessed for monitored natural attenuation. The desired final concentration of benzene is 5 ppb. Considering flushing and dilution only, how many years will it take to reduce the benzene concentration to 5 ppb? After several months of sampling and pilot tests onsite, it is found that, with flushing and natural biodegradation, the first-order decay rate is twice the rate of flushing alone. How many years will it take to achieve the cleanup goal considering both flushing and biodegradation? Additional hydrogeologic parameters are listed below.

- Groundwater seepage velocity = 7.5×10^{-5} cm/s
- Retardation factor = 1.24

SOLUTION:

(a) Calculate the number of pore volumes to flush the benzene:

$$PV = -R \ln\left(\frac{C_f}{C_i}\right) = -1.24 \ln\left(\frac{5}{15,000}\right) = 9.93$$

The time for groundwater to flow through the plume is

$$\tau = \frac{L}{v} = \frac{58 \text{ m}}{7.5 \times 10^{-5} \text{ cm/s}} \times \frac{100 \text{ cm}}{1 \text{ m}} = 7.73 \times 10^7 \text{ s} = 2.45 \text{ years}$$

The time to flush to 5 ppb is

$$T = (PV) \times \tau = 9.93 \times 2.45 \text{ years} = 24.4 \text{ years}$$

(b) To find the combined first-order decay constant, first find the decay constant for flushing alone:

$$k = \frac{1}{R\tau} = \frac{1}{(1.24)(2.45)} = 0.329 \text{ yr}^{-1}$$

With biodegradation, the rate doubles to $0.329 \text{ yr}^{-1} \times 2 = 0.658 \text{ yr}^{-1}$. Now calculate the time of natural attenuation considering the new decay rate:

$$C_f = C_i e^{-kT}$$

$$\ln\left(\frac{C_f}{C_i}\right) = -kT$$

$$T = \frac{\ln(C_f / C_i)}{-k} = \frac{\ln(5/15,000)}{-0.658} = 12.2 \text{ years}$$

This section on MNA describes the guiding principles for designing a monitoring strategy to evaluate this natural process. Chapter 8 will provide further details on how to evaluate long-term data collected from MNA sites.

6.6 SUMMARY

This chapter discussed common groundwater remediation techniques:

- Pump-and-treat
  - Pumping wells
  - Capture zones
  - Activated carbon
  - Air stripping
  - Advanced oxidation
  - Chemical precipitation

- *In situ* groundwater remediation
  - Air sparging
  - Bioremediation
  - Chemical oxidation
  - Monitored natural attenuation

Many of these techniques can be combined. For example, pump-and-treat can be used for hydraulic containment to protect a sensitive receptor, while conducting aggressive *in situ* bioremediation in "hot spots" at the same site.

6.7 PROBLEMS AND ACTIVITIES

6.1. Do an internet image search for groundwater pumping wells. Neatly draw, by hand or by using software, a groundwater pumping well based on your search that includes, at a minimum: the casing, the well cap, the screen, the filter (or filter pack), a submersible pump, aquifer drawdown, and aquifer cone of depression. Cite the source(s) of your information.

6.2. Describe in words and a sketch the difference between the terms drawdown (s) and static head (h).

6.3. Use the following information to estimate the groundwater extraction rate of a pumping well in an unconfined aquifer:
- Aquifer thickness = 9.1 m thick
- Well diameter = 4-inch (0.10 m) diameter
- Hydraulic conductivity of the aquifer = 16.3 m/d
- Steady-state drawdown = 0.61 m in a monitoring well which is 1.52 m from the pumping well
  = 0.36 m observed in a monitoring well which is 6.10 m from the pumping well

6.4. A water table (unconfined) aquifer is 12.2 m thick. Groundwater is being extracted from a 4-in (0.10 m) diameter fully penetrating well. The extraction rate is 0.15 m³/min. The aquifer is relatively sandy and has a hydraulic conductivity of 8.15 m/day. Steady-state drawdown of 1.85 m is observed in a monitoring well at 4.20 m from the pumping well. Determine
   a) The drawdown in the pumping well
   b) The radius of influence of the pumping well.

6.5. Use the following information to estimate the groundwater extraction rate of a fully-penetrating well in a confined aquifer:
- Aquifer thickness = 12.2 m
- Well diameter = 0.1 m
- Hydraulic conductivity = 1×10^{-2} cm/s
- Steady-state drawdowns: 1.5 m in a monitoring well 3.0 m from the pumping well and 1.2 m in a monitoring well 6.0 m from the pumping well

6.6. Make a plan view sketch of a moving contaminated plume and a pumping well capture zone that encompasses the plume. Label the capture zone's x- and y-axes, the groundwater flow direction, the stagnation point distance, and the full width at the well.

6.7. A groundwater extraction well is installed in an aquifer (hydraulic conductivity = 40.8 m/day, hydraulic gradient=0.015, and aquifer thickness=24.4 m). The design pumping rate is 3.15 L/s. Delineate the capture zone of this extraction well by specifying the following characteristic distances of the capture zone:

   a. the side-stream distance from the well to the envelope of the capture zone at the line of the pumping well,

   b. the down-stream distance from the well to stagnation point of the envelope, and

   c. the side-stream distance of the envelope far upstream of the pumping well.

6.8. A drum of formaldehyde has leaked into a sandy soil. The soil has a hydraulic conductivity of 6.4×10^{-4} m/s and a porosity of 0.36. The groundwater table is 2.5 m below grade and has a hydraulic gradient of 0.0018. The aquifer is 15 m thick. A single well used for hydraulic containment is proposed, using a pumping rate of 0.0090 m³/s. Calculate the full width of the capture zone far away from the well and at the well.

6.9. The results of a recent remedial investigation indicate that the groundwater plume has migrated 20 ft off-site. To prevent further off-site migration and to capture the off-site plume, a groundwater extraction well is to be installed at the property line. What would be the minimum groundwater extraction rate to contain the plume (aquifer porosity=0.350; aquifer thickness=60 ft; groundwater gradient=0.015; aquifer hydraulic conductivity= 1,000 gpd/ft²).

6.10. Groundwater contaminated with chlorobenzene is to be treated using granular activated carbon. What is the mass (kg) of GAC used per day?
   - C_{in}=500.0 mg/L
   - k=91 and 1/n=0.99, yielding a q in units of μg COC/g GAC
   - Q=1.0×10^6 gallons per day

6.11. Dewatering to lower the groundwater level for below-ground construction is often necessary. At a construction site, the contractor unexpectedly found that the extracted groundwater contained tetrachloroethylene (PCE). The PCE concentration of the groundwater has to be reduced before discharge. To avoid further delay of the tight construction schedule, off-the-shelf 55-gallon (200 L) activated carbon units are proposed to treat the groundwater. Using the following information to design an activated carbon treatment system (i.e., number of carbon units and configuration of flow):
   - the adsorption capacity is q=0.008 g PCE/g GAC
   - water flow rate= 190 L/min
   - diameter of carbon packing bed in each 55-gallon drum=0.5 m

- height of carbon packing bed in each 55-gallon drum = 1.0 m
- bulk density of GAC = 480 kg/m^3
- EBCT = 10 min minimum
- SLR = m^3 h^{-1} m^{-2} maximum

6.12. A packed-column air stripper is designed to reduce benzene concentration in the extracted groundwater. The concentration is to be reduced from 5 mg/L to 0.005 mg/L. Size the air stripper by determining the cross-sectional surface area, packing height, and the air flow rate. Use the following information in your calculations:

- Stripping factor = 5
- Temperature of the water = 25°C
- Henry's constant of benzene (dimensionless) = 0.23 @25°C
- Groundwater flow rate = 200 gpm
- Hydraulic loading rate = 20 gpm/ft^2
- $K_La = 0.02$/s
- Contaminant concentration in the influent air = 0.

6.13. A packed-column air stripper is used to reduce chloroethane (C_2H_5Cl) from the extracted groundwater from 3 mg/L to 30 µg/L. Henry's constant of chloroethane @ 20°C = 0.6; stripping factor = 6; operating temperature of the air stripper = 20°C; $K_{La} = 0.035$/s; column diameter = 2 m; and groundwater flow rate = 4 m^3/min. Determine

a) the required air flow rate
b) the hydraulic loading rate
c) the height of transfer unit
d) the required number of transfer units
e) the required packing height

6.14. UV and ozone treatment is selected to remove PCE from an extracted groundwater stream (PCE concentration = 1,250 ppb). A pilot study was conducted with one *ex situ* reactor designed for a hydraulic residence time of 3 minutes. This system could reduce PCE concentration from 1,250 ppb to 120 ppb. However, the discharge limit for PCE is 5 ppb. Assuming the reactors behave as ideal plug-flow reactors and that the reaction is of first-order, how many of these reactors in series would you recommend to use? What would be the final PCE concentration with the number of recommended reactors?

6.15. An air sparging well was installed into the plume of an aquifer. The injection air flow rate into the well is 5 ft^3/min. The height of the water column above the air injection point is 8.0 ft. The aquifer matrix consists mainly of medium sand. Determine the minimum air injection pressure required.

6.16. A biosparging well was installed into the plume of an aquifer impacted by petroleum hydrocarbons. The injection air flow rate into the well is 0.21 m^3/min. Assuming the oxygen transfer efficiency (OTE) is 9%, determine the rate of oxygen addition to the aquifer through the biosparging well. What would be the equivalent injection rate of water with a dissolved oxygen (DO) concentration of 9 mg/L?

6.17. A groundwater aquifer is impacted by gasoline. The average dissolved gasoline concentration of the groundwater samples is 25 mg/L. At this site, on average, 1 m³ of aquifer contains 340 g of gasoline. The porosity is 0.40. Assume gasoline has the same formula as heptane, C_7H_{16}. *In situ* bioremediation is being considered for aquifer restoration, using $(NH_4)_2SO_4$ and $Na_3PO_4 \cdot 12H_2O$ as the added N and P compounds.

Assuming no nutrients are available in the groundwater for bioremediation and the optimal molar C:N:P ratio has been determined to be 100:15:1, determine the mass of nutrients needed to support the biodegradation of gasoline, on a basis of 1 m³ of aquifer. If the plume is to be flushed with 100 pore volumes of oxygen- and nutrient-enriched water, what would be the required nutrient concentration in the water for reinjection?

6.18. You have been placed in charge of selecting locations for groundwater monitoring wells to monitor natural attenuation at the site shown in Figure 6.19. Your budget is limited to 7 wells. Show the location of each well and explain the reason for each location. Be careful not to place wells where buildings might be located.

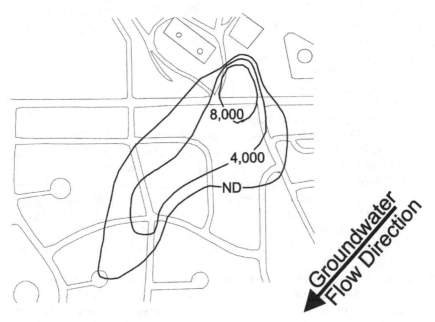

FIGURE 6.19 Plan view of a plume distribution. Concentrations are in units of ppb (Problem 6.18). (Modified from Wiedemeier et al. 2000).

6.19. A toluene plume of length 125 m at an average concentration of 25 mg/L is to be assessed for monitored natural attenuation. The desired final concentration of benzene is 1 mg/L. Considering flushing and dilution only, how many years will it take to reduce the toluene concentration? After several months of sampling and pilot tests onsite, it is found that, with flushing and natural biodegradation, the first-order decay rate is twice the rate of flushing alone. How many years will it take to achieve the cleanup goal considering both flushing and biodegradation? Additional hydrogeologic parameters are listed below.

- Groundwater seepage velocity $= 1.0 \times 10^{-4}$ cm/s
- Retardation factor $= 1.75$

REFERENCES

Davis, M. L., and Masten, S. J. (2013). *Principles of Environmental Engineering and Science.* McGraw-Hill Education, New York.

ITRC. (2005). *Overview of Groundwater Remediation Technologies for MTBE and TBA.* Interstate Technology & Regulatory Council, Washington, DC.

Javandel, I., and Tsang, C.-F. (1986). "Capture-zone Type Curves: A Tool for Aquifer Cleanup." *Groundwater*, 24(5), 616–625.

Johnson, R. C., Johnson, P. C., McWhorter, D. B., Hinchee, R. E., and Goodman, I. (1993). "An Overview of In Situ Air Sparging." *Ground Water Monitoring Review*, Fall, 127–135.

Kuo, J. (2014). *Practical Design Calculations for Groundwater and Soil Remediation.* CRC Press, Boca Raton, FL.

NJDEP. (2012). *Site Remediation Program Technical Guidance, Monitored Natural Attenuation Technical Guidance.* New Jersey Department of Environmental Protection, Trenton, NJ.

Ong, S. K., and Kolz, A. (2007). "Chemical Treatment Technologies." *Remediation Technologies for Soils and Groundwater*, A. Bhandari, R. Surampalli, P. Champagne, S. K. Ong, R. D. Tyagi, and I. Lo, eds., American Society of Civil Engineers, Reston, VA, 79–132.

Pope, D. F., Acree, S. D., Levine, H., Mangion, S., van Ee, J., Hurt, K., and Wilson, B. (2004). *Performance Monitoring of MNA Remedies for VOCs in Ground Water.* EPA/600/R-04/027. U.S. Environmental Protection Agency National Risk Management Research Laboratory, Cincinnati, OH.

U.S. EPA. (1991). *Site Characterization for Subsurface Remediation.* EPA/625/4-91/026. U.S. Environmental Protection Agency Office of Research and Development, Washington, DC.

U.S. EPA. (2001). *Groundwater Pump and Treat Systems: Summary of Selected Cost and Performance Information at Superfund-financed Sites.* EPA 542-R-01-021b. U.S. Environmental Protection Agency Office of Solid Waste and Emergency Response, Cincinnati, OH.

U.S. EPA. (2006). *Off-Gas Treatment Technologies for Soil Vapor Extraction Systems: State of the Practice.* EPA-542-R-05-028. U.S. Environmental Protection Agency Office of Superfund Remediation and Technology Innovation, Washington, DC.

U.S. EPA. (2012a). *A Citizen's Guide to Activated Carbon Treatment.* EPA 542-F-12-001. U.S. Environmental Protection Agency Office of Solid Waste and Emergency Response, Washington, DC.

U.S. EPA. (2012b). *A Citizen's Guide to Air Stripping.* EPA 542-F-12-002. U.S. Environmental Protection Agency Office of Solid Waste and Emergency Response, Washington, DC.

U.S. EPA. (2012c). *A Citizen's Guide to Monitored Natural Attenuation.* EPA 542-F-12-014. U.S. Environmental Protection Agency Office of Solid Waste and Emergency Response, Washington, DC.

U.S. EPA. (2017a). *How To Evaluate Alternative Cleanup Technologies For Underground Storage Tank Sites, Chapter VII Air Sparging.* EPA 510-B-17-003. U.S. Environmental Protection Agency Office of Land and Emergency Management, Washington, DC.

U.S. EPA. (2017b). *How to Evaluate Alternative Cleanup Technologies for Underground Storage Tank Sites, Chapter XII Enhanced Aerobic Bioremediation.* EPA 510-B-17-003. U.S. Environmental Protection Agency Land and Emergency Management, Washington, DC.

U.S. EPA. (2017c). *How to Evaluate Alternative Cleanup Technologies for Underground Storage Tank Sites, Chapter IX Monitored Natural Attenuation.* EPA 510-B-17-003. U.S. Environmental Protection Agency Land and Emergency Management, Washington, DC.

Wiedemeier, T. H., Lucas, M. A., and Hass, P. E. (2000). *Designing Monitoring Programs to Effectively Evaluate the Performance of Natural Attenuation.* Air Force Center for Environmental Excellence Technology Transfer Division, San Antonio, TX.

Wisconsin DNR. (2014). *Guidance on Natural Attenuation for Petroleum Releases. PUB-RR-614.* Wisconsin Department of Natural Resources, Madison, WI.

7 Off-Gas Treatment

7.1 INTRODUCTION

The remediation of impacted soil and groundwater often results in the transfer of organic chemicals of concern (COCs) from soil and groundwater into the air. The air stream containing organic COCs, known as off-gas, usually needs to be treated before being released to the atmosphere. The development and implementation of an air emission control strategy should be an integral part of the overall remediation program. Air emission control can be expensive, and it may affect the cost-effectiveness of a specific remedial alternative.

COCs in off-gases are necessarily volatile organic compounds (VOCs). Common sources of VOC-laden off-gas from soil and groundwater remediation activities include soil vapor extraction, soil washing, solidification/stabilization, air sparging, biosparging, air stripping, and bioremediation. This chapter covers some design calculations for commonly used off-gas treatment technologies, including activated carbon adsorption, thermal oxidation, and catalytic oxidation.

7.2 ACTIVATED CARBON ADSORPTION

Activated carbon adsorption is the most commonly used air pollution control processes for VOC emissions (U.S. EPA 2006). The process is very effective in removing a wide range of VOCs. The most common form of activated carbon for air emission control is granular activated carbon (GAC).

Granular activated carbon (GAC) has a look and consistency of crushed charcoal (see Figure 6.6). It is found in everyday products such as household water filters for drinking water and fish aquariums. GAC is manufactured from combusted products like wood and coconut shells. Through this process, the GAC granules become extremely porous and have a large surface area available for adsorption. The porous structure of a GAC grain can be seen as a close-up in Figure 6.6. The pores take the shape of a tree root, which provides a large surface area for organic molecules to adsorb. Adsorption is the action of a COC molecule traveling from air or water and accumulating on a solid surface (the adsorbent). The adsorbed molecule is called the adsorbate. For use in remediation, this granular material is placed inside tanks, or vessels, properly designed and built to handle the fluid flow and COC concentrations (Figure 7.1). This chapter focuses on GAC adsorption for air flows, while Chapter 6 introduced GAC adsorption for liquid flows. The main concepts are the same, though activated carbon grains for vapor and water are manufactured differently.

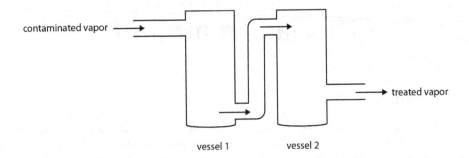

contaminated vapor →

treated vapor

vessel 1   vessel 2

FIGURE 7.1   A general schematic of two GAC vessels with contaminated air flow through, with treated air exiting the second vessel.

7.2.1   Adsorption Isotherm and Adsorption Capacity

The adsorption capacity of GAC depends on the characteristics of GAC, characteristics of VOCs and their concentration, temperature, and presence of other species competing for adsorption. At a given temperature, a relationship exists between the mass of the VOC adsorbed per unit mass GAC to the concentration (or partial pressure) of VOC in the waste air stream. For most of the VOCs, the adsorption isotherms can be fitted well by a power curve, in particular the Freundlich isotherm (also see Eq. 6.12):

$$q = a\left(P_{VOC}\right)^m \tag{7.1}$$

where
   q = equilibrium adsorption capacity, mass of VOC/mass of GAC
   P_{VOC} = partial pressure of VOC in the waste air stream, psi
   a, m = empirical constants

Empirical constants of the Freundlich isotherms for selected VOCs are listed in Table 7.1. It should be noted that the values of these empirical constants are for a specific type of GAC only and should not be used outside the specified range. In addition, for the constants in Table 7.1, q results specifically in units of lb COC/lb GAC. Other sources of Freundlich constants may result in units of mg COC/kg GAC or other variations of mass units.

The actual adsorption capacity in field applications would be lower because the adsorption isotherms are usually developed in a laboratory setting, without the presence of other compounds that would compete for the adsorption sites. Normally, design engineers take about 50% of this theoretical value as the design adsorption capacity as a factor of safety. Therefore,

$$q_{design} = \left(50\%\right)\left(q_{theoretical}\right) \tag{7.2}$$

Keep in mind that, by definition from Eq. 7.1,

$$q = \frac{Mass\,VOC}{Mass\,GAC} \qquad (7.3)$$

So after q is established for a specific vessel and mass of GAC, one can solve for the mass of VOC that the vessel can retain.

TABLE 7.1

Empirical Constants for Selected Adsorption Isotherms*

Compounds	Adsorption Temperature (°F)	a	M	Range of P_{VOC} (psi)
Benzene	77	0.597	0.176	0.0001–0.05
Toluene	77	0.551	0.110	0.0001–0.05
m-Xylene	77	0.708	0.113	0.0001–0.001
m-Xylene	77	0.527	0.0703	0.001–0.05
Phenol	104	0.855	0.153	0.0001–0.03
Chlorobenzene	77	1.05	0.188	0.0001–0.01
Cyclohexane	100	0.508	0.210	0.0001–0.05
Dichloroethane	77	0.976	0.281	0.0001–0.04
Trichloroethane	77	1.06	0.161	0.0001–0.04
Vinyl chloride	100	0.20	0.477	0.0001–0.05
Acrylonitrile	100	0.935	0.424	0.0001–0.05
Acetone	100	0.412	0.389	0.0001–0.05

*The resulting q is in units of lb COC/lb GAC
Source: U.S. EPA 1991.

Example 7.1: Determine the Capacity of a Vapor Phase GAC Adsorber

The off-gas from a soil vapor extraction project is to be treated by GAC adsorbers. The chlorobenzene concentration in the off-gas is 500 ppmV. The flow rate out of the vacuum pump is 150 ft³/min (cfm), (4.25 m³/min), and the temperature of the air is ambient. Two 1,000-lb (450 kg) activated carbon adsorbers are proposed. Determine the maximum mass of chlorobenzene that can be held by each GAC adsorber before exhausted. Use the isotherm data in Table 7.1.

SOLUTION:

Convert the chlorobenzene concentration from ppmV to psi as:

$P_{VOC} = 500$ ppmV $= 500 \times 10^{-6}$ atm $= 5.0 \times 10^{-4}$ atm

$= (5.0 \times 10^{-4}$ atm$)(14.7$ psi/atm$) = 0.00735$ psi

Obtain the empirical constants for the adsorption isotherm from Table 7.1 and then apply Eq. 7.1 to determine the equilibrium adsorption capacity as:

$$q_{theoretical} = a(P_{VOC})^m = (1.05)(0.00735)^{0.188} = 0.417 \frac{lb\,chlorobenzene}{lb\,GAC}$$

The design adsorption capacity can be found by using Eq. 7.2 as:

$$q_{design} = (50\%)(q_{theoretical}) = (50\%)(0.417) = 0.208 \frac{lb\,chlorobenzene}{lb\,GAC}$$

Use Eq. 7.3 to calculate the mass of chlorobenzene that can be retained by an adsorber before the GAC becomes exhausted

mass of VOC $= (\text{mass of GAC})(q_{design})$

$= (1,000\ lb\ GAC/unit)(0.208\ lb\ chlorobenzene\ /\ lb\ GAC) = 208\ lb\ chlorobenzene/unit.$

DISCUSSION:

1. The design capacity is found to be 0.208 lb chlorobenzene/lb GAC, which is equal to 0.208 kg chlorobenzene/kg GAC.
2. The adsorption capacity of vapor-phase GAC is typically of the order of magnitude of 0.1 lb/lb (or 0.1 kg/kg), which is much higher than the adsorption capacity of liquid-phase GAC, typically of the order of magnitude of 0.01 lb/lb (or 0.01 kg/kg).
3. Care should be taken to use matching units for P_{voc} and q in the adsorption isotherms, and to use the applicable range of P_{VOC}.
4. The influent COC concentration in the air stream, not the effluent concentration, should be used in the adsorption isotherms to determine the adsorption capacity.

7.2.2 DESIGN OF GAC ADSORBERS

To achieve efficient adsorption, the air flow rate through the activated carbon should be kept as low as possible. The practical design velocity, also known as air flow velocity, is often ≤60 ft/min (≤18 m/min), and 100 ft/min (30 m/min) is considered as the maximum value. This design parameter can be used to determine the cross-sectional area of the GAC adsorbers (A_{GAC}):

$$A_{GAC} = \frac{Q}{Air\,Flow\,Velocity} \tag{7.4}$$

where Q is the influent air flow rate. The design height of the adsorber is normally 2 ft (0.7 m) or deeper to provide a sufficiently large mass transfer zone for adsorption.

Example 7.2: Required Cross-sectional Area of GAC Adsorbers

Referring to the remediation project described in Example 7.1, the 1,000-lb GAC units are out of stock. To avoid delay of remediation, off-the-shelf 55-gallon activated carbon units are proposed on an interim basis. The type of GAC in the

55-gallon units is same as that in the 1,000-lb units, so the isotherm data are the same. The vendor also provided the following information with regard to the units:

- Diameter of carbon packing bed in each 55-gallon drum = 1.5 ft
- Height of carbon packing bed in each 55-gallon drum = 3 ft
- Bulk density of the activated carbon = 28 lb/ft³
- Design velocity = 50 ft/min

Determine (a) the mass of activated carbon in each 55-gallon unit, (b) the mass of chlorobenzene that each unit can remove before exhausted, and (c) the minimum number of the 55-gallon units needed.

SOLUTION:

(a) Volume of the activated carbon inside a 55-gallon drum:

$$V_{drum} = \frac{\pi D^2}{4} H = \frac{\pi (1.5)^2}{4} (3) = 5.30 \, ft^3$$

Mass of activated carbon inside a 55-gallon drum:

$$M_{GAC} = \left(5.3 \, ft^3\right) \left(28 \frac{lb}{ft^3}\right) = 148 \, lb$$

(b) Mass of chlorobenzene that can be retained by a drum before the GAC becomes exhausted:
= (mass of the GAC)(design adsorption capacity from Example 7.1)
= (148 lb/drum)(0.208 lb chlorobenzene/lb GAC) = 31.0 lb chlorobenzene/drum.

(c) Using the design air flow velocity of 50 ft/min, the required cross-sectional area for GAC adsorption can be found using Eq. 7.4:

$$A_{GAC,req'd} = \frac{Q}{air \; flow \; velocity} = \frac{150 \, ft^3/min}{50 \, ft/min} = 3.0 \, ft^2$$

If the adsorption vessel were tailor-made, then a system with a cross-sectional area of 3.0 ft² would do the job. However, since the off-the-shelf 55-gallon drums are to be used, we need to determine the number of drums that will provide the required cross-sectional area.

Area of the activated carbon inside a 55-gallon drum:

$$A_{drum} = \frac{\pi D^2}{4} = \frac{\pi (1.5)^2}{4} = 1.77 \frac{ft^2}{drum}$$

The cross-sectional area of each drum is 1.77 ft², less than the required 3.0 ft².

Next, determine the number of drums in parallel to meet the required flow rate and air flow velocity:

$$Number \, of \, drums = \frac{A_{GAC}}{A_{drum}} = \frac{3.0}{1.77} = 1.7 \, drums, \; round \; up \; to \; 2 \; drums$$

So, use 2 drums in parallel to provide the required cross-sectional area. The total cross-sectional area of 2 drums is equal to 3.54 ft² (=1.77×2).

Discussion: The minimum number of 55-gallon drums for this project is 2 to meet the air flow velocity requirement.

The actual number of GAC vessels should be more than the design in the previous example to meet the monitoring requirements or the desirable frequency of change-out. GAC adsorbers are usually set up at least two in series (see Figure 7.2). When two adsorbers are arranged in series, the monitoring point is located at the effluent of the first adsorber. A high effluent concentration from the first adsorber indicates that this adsorber is reaching its capacity. The first adsorber is then taken off-line, and the second adsorber is shifted to be the first adsorber. Consequently, the capacity of both adsorbers can be fully utilized, and the compliance requirements can also be met. If there are two parallel streams of adsorbers, one stream can be taken off-line for regeneration or maintenance, while the other parallel stream can continue to operate.

The COC removal rate by a GAC adsorber ($R_{removal}$) can be calculated by using the following formula:

$$R_{removal} = \left(G_{in} - G_{out}\right) \qquad (7.5)$$

In practical applications, the effluent concentration (G_{out}) is kept below the discharge limit, which is often very low. Therefore, for a factor of safety, the term G_{out} can be

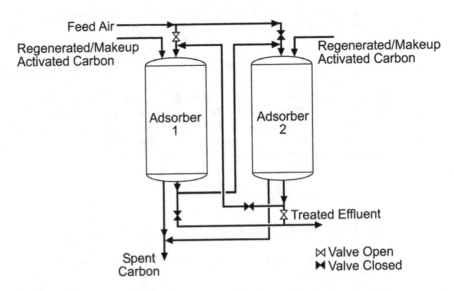

FIGURE 7.2 Arrangement of GAC vessels in series. (Source: U.S. EPA 2006.)

deleted from Eq. 7.5 in design. The mass removal rate is then the same as the mass loading rate ($R_{loading}$):

$$R_{removal} \approx R_{loading} = (G_{in})Q \qquad (7.6)$$

The mass loading rate is nothing but the product of the air flow rate and the COC concentration. As mentioned in Chapter 2, the contaminant concentration in the air is often expressed in ppmV or ppbV. In the mass loading rate calculation, the concentration has to be converted into mass concentration units using Eq. 2.5 or Eq. 2.6. Because this is an operational issue, it is covered in Section 8.2.1.

Once the activated carbon reaches its capacity, it should be regenerated or disposed of. The time interval between two regenerations or the expected service life of a fresh batch of GAC can be found by dividing the capacity of GAC with the COC removal rate ($R_{removal}$) as:

$$T = \frac{M_{removed}}{R_{removal}} \qquad (7.7)$$

Example 7.3: Determine the Change-out (or Regeneration) Frequency of the GAC Adsorbers

Referring to the remediation project described in Example 7.2, the capacity of each GAC unit is 31.0 lb chlorobenzene/drum. If the mass removal rate is 0.042 lb/min, determine the service life of the two 55-gallon GAC units.

SOLUTION:

Use Eq. 7.9 to determine the service life of the two drums:

$$T = \frac{M_{removed}}{R_{removal}} = \frac{(2)(31.0\,lb)}{0.042\,lb/min} = 1,480 \text{ min} = 24.7 \text{ hours}$$

DISCUSSION:

1. Although two drums in parallel can provide a sufficient cross-sectional area for adequate air flow velocity, the relatively high contaminant concentration makes the service life of the two 55-gallon drums unacceptably short.
2. A 55-gallon activated carbon drum normally costs several hundred dollars. In this example, two drums last about one day. The labor and disposal costs should also be added, and it makes this option prohibitive. A GAC system with on-site regeneration or other treatment alternatives should be considered.

7.3 THERMAL OXIDATION

Thermal processes are also commonly used to treat VOC-laden air. Thermal oxidation, catalytic oxidation, and internal combustion (IC) engines are popular thermal

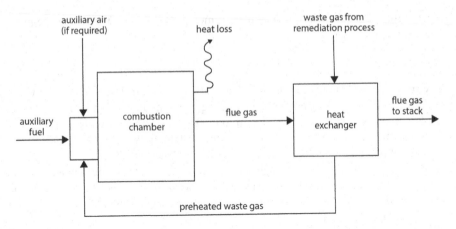

FIGURE 7.3 Schematic of flows through a thermal oxidizer. (Modified from U.S. EPA 1992.)

processes for these applications. These processes, which can be referred to by the words oxidation, incineration, and combustion, have the purpose of destroying organic compounds. In other words, organic molecules are degraded into smaller and nontoxic molecules such as water and carbon dioxide.

The key components of thermal treatment system design are the "three T's": combustion temperature, residence time (also called "dwell time"), and turbulence. They affect the size of a reactor and its VOC destruction efficiency. For example, to achieve good thermal destruction, the VOC-laden air should be held inside a thermal oxidizer for a sufficient residence time (normally 0.3–1.0 seconds) at a high temperature, at least 100°F above the auto-ignition temperatures of the COCs in the VOC-laden gas stream. In addition, sufficient turbulence must be maintained in the oxidizer to assure good mixing and complete combustion of the COCs. Other important parameters to be considered include the heating value of the influent and the requirements of auxiliary fuel and supplementary air.

A discussion on the combustion basics for thermal oxidation will be presented here and is essentially applicable to other thermal processes.

Figure 7.3 gives an overview of a thermal oxidation process. The waste gas from the remediation process, in other words, the extracted soil vapor, first enters the heat exchanger. The heat exchanger heats the gas to be combusted. This gas, as it is laden with organic compounds, also serves as fuel for combustion (see Section 7.3.2). The combusted gas is called flue gas and leaves the combustion chamber, passing again through the heat exchanger. In the heat exchanger, some of the treated gas is discharged to the atmosphere through the stack, and some is recirculated back into the combustion chamber. On the left side in Figure 7.3 we also see auxiliary fuel and auxiliary air entering the combustion chamber. Auxiliary fuel is needed for combustion in case the waste gas does not provide enough heating value for combustion. Auxiliary air (Section 7.3.3) is atmospheric supplementary air added to the combustion chamber in case the gas does not have enough oxygen to promote combustion.

7.3.1 AIR FLOW RATE VERSUS TEMPERATURE AND PRESSURE

Because air expands and contracts according to temperature and pressure, we will first discuss the concepts of standard air flow rate and actual air flow rate. This distinction is important when comparing flows throughout a system or across systems.

The volumetric air flow rate is commonly expressed in ft³/min, that is, cubic feet per minute (cfm), or m³/min. Since the volumetric flow rate of an air stream is a function of temperature and pressure, and the air stream undergoes zones of different temperatures and pressures in a thermal process, the air flow rate that is measured with instruments is the actual cfm (acfm). The unit of acfm refers to the volumetric flow rate under the actual temperature and pressure. The standard cfm (scfm) is the flow rate at standard conditions, that is, standard temperature and pressure. Standard pressure is always atmospheric pressure, that is, 1 atm or 14.7 psi or 101.3 kPa or 760 mm-Hg. However, the definition of the standard temperature is not universal. For the EPA, the standard temperature is $T = 77°F$ (25°C). But 32°F (0°C) or 68°F (20°C) are also commonly used in technical articles as the standard temperature. One should follow the regulatory requirements and use the appropriate reference temperature for a specific project. A standard temperature of 77°F (25°C) is used in this chapter, unless otherwise specified.

Conversions between actual and standard flow rates for a given air stream can be made assuming that the Ideal Gas Law is valid. Consider the Ideal Gas Law applied to a point in a vapor treatment system where the air stream is measured (the actual measurement ("a") and needs to be converted to a standard air stream flow "s." The Ideal Gas Law (see Eq. 2.7a) applies in both scenarios:

$$P_a V_a = n_a R T_a \qquad (7.8a)$$

$$P_s V_s = n_s R T_s \qquad (7.8b)$$

Note that R is the universal gas constant and is the same in both equations. The same number of moles is passing through that location in the system, so n_a and n_s are equal. We can solve Eqs. 7.8a and 7.8b for nR and make them equal to each other:

$$\frac{P_a V_a}{T_a} = \frac{P_s V_s}{T_s} \qquad (7.9)$$

We can rearrange the equation above to solve for the ratio V_a/V_s:

$$\frac{V_a}{V_s} = \frac{P_s}{P_a} \times \frac{T_a}{T_s} \qquad (7.10)$$

An air stream flow is the volumetric flow rate, and the parcel of volume V can be substituted with Q:

$$\frac{Q_a}{Q_s} = \frac{P_s}{P_a} \times \frac{T_a}{T_s} \qquad (7.11)$$

Remember that in the Ideal Gas Law, the absolute pressure is used, and the temperature is in Kelvin (for SI units, so 273+°C) or Rankine (for USCS, so 460+°F). The following example shows an application of the relationship between the actual flow rate and the standard flow rate.

Example 7.4: Conversion between Actual and Standard Air Flow Rates

A thermal oxidizer was used to treat the off-gas from a soil vapor extraction system. To achieve the required COC removal efficiency, the oxidizer was operated at 1,400°F (760°C). The measured actual flow rate immediately leaving the oxidizer chamber is 550 ft³/min (15.6 m³/min) at nearly atmospheric pressure. The air flow then exits through a 4-in. (10 cm) diameter stack, and the temperature has decreased to 200°F (93°C). (a) What would be the flow rate exiting the oxidizer chamber, expressed in scfm and in standard m³/min? (b) What is the air flow velocity leaving the stack?

SOLUTION:

(a) Use Eq. 7.11 to convert actual flow rate to standard flow rate as:

$$Q_s = Q_a \times \frac{P_a}{P_s} \times \frac{T_s}{T_a} = 550 \frac{ft^3}{min} \times \frac{1\,atm}{1\,atm} \times \frac{460+77}{460+1,400} = 158.8\,scfm$$

$$Q_s = Q_a \times \frac{P_a}{P_s} \times \frac{T_s}{T_a} = 15.6 \frac{m^3}{min} \times \frac{1\,atm}{1\,atm} \times \frac{273+25}{273+760} = 4.5\,m^3/min$$

(b) Use Eq. 7.11 to determine the flow rate from the stack:

$$Q_a = Q_s \times \frac{P_s}{P_a} \times \frac{T_a}{T_s} = 158.8 \frac{ft^3}{min} \times \frac{1\,atm}{1\,atm} \times \frac{460+200}{460+77} = 195.2\,cfm$$

$$Q_a = Q_s \times \frac{P_s}{P_a} \times \frac{T_a}{T_s} = 4.5 \frac{m^3}{min} \times \frac{1\,atm}{1\,atm} \times \frac{273+93}{273+25} = 5.5\,m^3/min$$

The discharge velocity is

$$v = \frac{Q}{A}$$

$$A = \frac{\pi D^2}{4} = \frac{\pi \left(\frac{4}{12}ft\right)^2}{4} = 0.0873\,ft^2 = \frac{\pi (0.10\,m)^2}{4} = 0.0079\,m^2$$

$$v = \frac{195.2\,ft^3/m}{0.0873\,ft^2} = 2,240 \frac{ft}{min} = \frac{5.5\,m^3/min}{0.0079\,m^2} = 696 \frac{m}{min}$$

Discussion: If the actual flow rate at one temperature is known, it can be used to determine the flow rate at another location and another temperature and pressure by using the following formula:

$$\frac{Q_1}{Q_2} = \frac{P_2}{P_2} \times \frac{T_1}{T_s}$$

In this example, if we consider the exit of the chamber to be location 1 and the exit of the stack to be location 2, we can find the stack flow rate as:

$$Q_2 = Q_1 \times \frac{P_1}{P_2} \times \frac{T_2}{T_1} = 550 \frac{ft^3}{min} \times \frac{1\,atm}{1\,atm} \times \frac{460+200}{460+1400} = 195.2\,cfm$$

$$= 15.6 \frac{m^3}{min} \times \frac{1\,atm}{1\,atm} \times \frac{273+93}{273+760} = 5.5\,m^3/min$$

7.3.2 Dilution and Auxiliary Air

Some waste air streams contain enough organic compounds to sustain burning, which means that purchasing additional auxiliary fuels is not necessary. That is why oxidation/incineration is favorable for treating air with high organic concentrations.

However, for safety concerns, care must be taken to not let the air stream reach flammability levels. Generally, the concentration of flammable vapors into a thermal oxidizer is limited by insurance companies to 25% of the lower explosive limit (LEL). The LEL of a gas is the percent volume concentration above which the vapor may become flammable under an ignition source, causing a fire or explosion concern.

Table 7.2 shows LELs for common gases, along with their upper explosive limits (UELs). The range between the LEL and the UEL is called the flammable range, which must be avoided. Vapor concentrations up to 40–50% of the LEL may be permissible if online monitoring of VOC concentrations and automatic process controls and shutdowns are employed.

When the off-gas has a VOC content larger than 25% of the LEL (i.e., in most of the initial stages of SVE-based cleanup projects), dilution air must be used to lower the COC concentration to below 25% of its LEL prior to oxidation, as shown in the schematic in Figure 7.4.

If the waste air stream has a low oxygen content (below 13–16%), then auxiliary air would also be used to raise the oxygen level to ensure flame stability of the burner. If the exact composition of the waste air stream is known, one can determine the stoichiometric amount of air (oxygen) for complete combustion. In general practice, excess air is added to ensure complete combustion.

7.3.3 Volume of the Combustion Chamber

The total influent to an oxidizer is the sum of the waste air (Q_w), dilution air (and/or auxiliary air, Q_d), and the supplementary (auxiliary) fuel (Q_{sf}), and if necessary, it can be determined by the following equation:

$$Q_{inf} = Q_w + Q_d + Q_{sf} \tag{7.12}$$

where Q_{inf}= the total influent flow rate.

TABLE 7.2

Lower Explosive Limits (LELs) and Upper Explosive Limits (UELs) of Common Gases

Compounds	LEL (% by Volume)	UEL (% by Volume)
Acetaldehyde	4	36[a]
Benzene	1.3	7[a]
Carbon monoxide	12.5	74.2
Ethane	3.1	12.4[a]
Ethylene	2.7	28.6
Gasoline (variable)	1.4–1.5	7.4–7.6
Heptane	1	6.7[a]
Hexane	1.2	7.4[a]
Methane	5	15
Methyl alcohol	6.7	36.5
Methyl ethyl ketone	1.8	9.5
Pentane	1.4	7.8
Propane	2.1	10.1
Propylene	2	11.1
Toluene	1.3	7
Vinyl chloride	4	21.7
Xylene	1	6

Sources: [a] U.S. EPA 1991. All others Lee and Huffmann (2007).

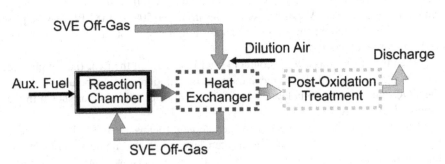

Source: Modified from U.S. EPA 2002
Processes in dashed boxes are not applicable to all oxidation systems.

FIGURE 7.4 Thermal oxidation schematic using dilution air. (Source: U.S. EPA 2006.)

In most cases, one can assume that the flow rate of the combined gas stream, Q_{inf}, entering the combustion chamber is approximately equal to the flue gas leaving the combustion chamber at standard conditions, Q_{fg}. The volume change across the oxidation chamber, due to combustion of VOC and supplementary fuel, is assumed to be small. This is especially true for dilute VOC streams from soil and groundwater remediation.

TABLE 7.3

Typical Thermal Oxidation System Design Values

Required Destruction Efficiency (%)	Nonhalogenated Compounds		Halogenated Compounds	
	Combustion Temperature	Residence Time (sec)	Combustion Temperature	Residence Time (sec)
98	1,600°F (871°C)	0.75	2,000°F (1,093°C)	1.0
99	1,800°F (982°C)	0.75	2,200°F (1,204°C)	1.0

Source: U.S. EPA 1991.

The flue gas flow rate in actual conditions can be determined from Eq. 7.11 (actual flow rate to standard flow rate).

The volume V_c of the combustion chamber is the product of the residence time τ and $Q_{fg,a}$, and can be found using the following equation:

$$V_c = Q_{fg,a}\tau \times 1.05 \tag{7.13}$$

The equation is nothing but "volume = flow rate × residence time." The safety factor 1.05 is an industrial practice to account for minor fluctuations in the flow rate. Table 7.3 tabulates the typical thermal oxidation system design values.

Example 7.5: Determine the Size of a Thermal Oxidizer

An off-gas stream ($Q = 210$ scfm = 6 standard m³/min) from a soil vapor extraction system containing 800 ppmV of xylenes is to be treated using a thermal oxidizer with a recuperative heat exchanger. Methane is added at a flow rate of 2.2 scfm (0.062 standard m³/min) as supplementary fuel to support combustion and no auxiliary air is needed. The combustion temperature is set at 1,800°F (982°C) to achieve a xylene destruction efficiency of 99% or higher. Determine the size of the thermal oxidizer.

SOLUTION:

(a) Use Eq. 7.12 to determine the flue gas flow rate at standard conditions:

$$Q_{fg} \sim Q_{inf} = Q_w + Q_d + Q_{sf} = 210 + 0 + 2.2$$

$$= 212.2 \text{ scfm} \left(= 6 + 0 + 0.062 = 6.06 \text{ m}^3/\text{min}\right)$$

(b) Use Eq. 7.8 to determine the flue gas flow rate at actual conditions:

$$Q_{fg,a} = Q_{fg,s} \times \frac{P_{fg,s}}{P_{fg,a}} \times \frac{T_{fg,a}}{T_{fg,s}} = 212.2 \frac{\text{ft}^3}{\text{min}} \times \frac{1\text{atm}}{1\text{atm}} \times \frac{460+1,800}{460+77} = 893 \text{ acfm}$$

$$= 6.06 \frac{\text{m}^3}{\text{min}} \times \frac{1\text{atm}}{1\text{atm}} \times \frac{273+982}{273+25} = 25.5 \text{ m}^3/\text{min}$$

From Table 7.3, the required residence time is 0.75 seconds. Use Eq. 7.13 to determine the size of the combustion chamber as:

$$V_c = Q_{fg,s}\tau \times 1.05 = 893\frac{ft^3}{min} \times 0.75\,s \times 1.05 \times \frac{1\,min}{60\,s} = 11.7\,ft^3$$

$$= 25.5\frac{m^3}{min} \times 0.75\,s \times 1.05 \times \frac{1\,min}{60\,s} = 0.33\,m^3$$

7.4 CATALYTIC OXIDATION

Catalytic oxidation, also known as catalytic incineration, is another commonly applied combustion technology for treating VOC-laden air. With the presence of a precious or base metal catalyst (Figure 7.5), the combustion temperature is generally between 600 and 1,200°F (315–650°C), which is lower than that of thermal oxidation systems.

For catalytic oxidation, the "three T's" (temperature, residence time, and turbulence) are still the important design parameters. Also, the type of catalyst has a significant effect on the system's performance and cost.

The concentration of flammable vapors to a catalytic oxidizer is generally limited to 20% of the LEL, which is lower than that for thermal oxidation. That is because higher VOC concentrations, and therefore higher temperatures, will damage the catalyst. Therefore, dilution air must be used to lower the COC concentration to below 20% of its LEL.

The total air stream influent to a catalyst bed is the sum of the waste air, dilution air (and/or auxiliary air), and supplementary fuel, if necessary. It can be determined from Eq. 7.12 for thermal oxidation.

In most of the cases, one can assume that the flow rate of the combined gas stream entering the catalyst, Q_{inf}, is approximately equal to the flue gas leaving the catalyst at standard conditions, Q_{fg}. The flue gas flow rate of actual conditions can be determined from Eq. 7.11.

Because of the short residence time in the catalyst bed, the parameter space velocity is commonly used to relate the volumetric air flow rate and the volume of the catalyst bed. The space velocity is defined as the volumetric flow rate of the VOC-laden air entering the catalyst bed divided by the volume of the catalyst bed. It is the

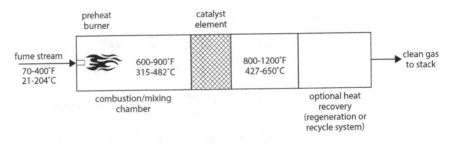

FIGURE 7.5 Schematic of a catalytic oxidizer. (Modified from U.S. EPA 1972.)

TABLE 7.4

Typical Design Parameters for Catalytic Incineration

Desired Destruction Efficiency (%)	Temperature at Catalyst Bed Inlet	Temperature at Catalyst Bed Outlet	Space Velocity (hr⁻¹)	
			Base Metal	Precious Metal
95	600°F (315°C)	1,000–1,200°F (540–650°C)	10,000–15,000	30,000–40,000

Source: U.S. EPA 1991.

inverse of residence time. Table 7.4 provides the typical design parameters for catalytic incinerators. It should be noted here that the flow rate used in the space velocity calculation is based on the influent gas flow rate at standard conditions, not that of the catalyst bed or the bed effluent.

The volume of the catalyst can be determined by:

$$V_{cat} = \frac{60Q_{inf}}{SV} \tag{7.14}$$

where V_{cat} is the volume of the catalyst bed, Q_{inf} is the total influent flow rate to the catalyst bed, in standard flow rate per minute, and SV is the space velocity, per hour.

Example 7.6: Determine the Size of the Catalyst Bed

Referring to the remediation project described in Example 7.5, an off-gas stream ($Q=210$ scfm$=6$ standard m³/min) containing 800 ppmV of xylenes is to be treated by a catalytic incinerator with a recuperative heat exchanger. The design space velocity is 12,000/hour. Determine the size of the catalyst bed.

SOLUTION:

(a) From Example 7.5 the flue gas flow rate at standard conditions:

$$Q_{fg} \sim Q_{inf} = Q_w + Q_d + Q_{sf} = 210+0+2.2 = 212.2 \text{ scfm}\left(=6+0+0.062=6.06 \text{ m}^3/\text{min}\right)$$

(b) With a space velocity of 12,000/hour, use Eq. 7.14 to determine the size of the catalyst bed:

$$V_{cat} = \frac{60Q_{inf}}{SV} = \frac{(60)(212.2\text{scfm})}{12,000} = 1.1\text{ft}^3 = \frac{(60)(6.06 \text{ m}^3/\text{min})}{12,000} = 0.03 \text{ m}^3$$

7.5   SUMMARY

This chapter covered three common off-gas treatment technologies:

- Activated carbon adsorption
- Thermal oxidation
- Catalytic oxidation

In addition, the following important concepts were described: conversion of actual to standard gas flow rates and lower and upper explosive limits (LEL and UEL).

7.6   PROBLEMS AND ACTIVITIES

7.1.   The discharge from a soil vapor extraction system has a concentration of 200 ppmV of acetone. This discharge is to be treated using GAC adsorbers. The flow rate from the vacuum pump that extracted soil vapor is 150 cfm, and the temperature of the air is ambient. Two 1,000-lb activated carbon adsorbers are proposed. (a) Determine the saturation mass of acetone that can be held by each GAC adsorber. Use the isotherm data in Table 7.1. (b) Compare this problem and its answer to that of Example 7.1 and discuss the appropriateness of using GAC to treat acetone.

7.2.   The discharge of a soil vapor extraction system flows at 300 cfm and contains acetone. Isotherm data indicate that 0.020 lb of acetone are removed per lb of GAC. 1,000-lb GAC units would be ideal to use in this scenario, but they are out of stock. To avoid delay of remediation, off-the-shelf 55-gallon GAC drums are proposed on a temporary basis. Information on each 55-gallon GAC unit is the following:

- Diameter of carbon packing bed in each 55-gallon drum = 1.5 ft
- Height of carbon packing bed in each 55-gallon drum = 3 ft
- Bulk density of the activated carbon = 28 lb/ft³
- Design velocity = 60 ft/min

Determine (a) the mass of chlorobenzene that each unit can remove before being exhausted, and (b) the minimum number of 55-gallon drums needed, considering that each stream must have two units in series.

7.3.   Soil venting, coupled with GAC for off-gas treatment, is considered as a remedial alternative for a site impacted by TPH. Two off-the-shelf GAC adsorbers are to be used in series (GAC volume = 1 m³/unit, GAC bulk density = 500 kg/m³, adsorption capacity of carbon = 0.1 kg TPH/kg GAC, TPH removal efficiency is 95%). If the TPH loading rate to the GAC system is 2 kg/day, estimate

(a) the TPH emission rate to the atmosphere, in kg/day.

(b) the breakthrough time if the system is running continuously and the influent is at this TPH loading rate.

7.4.   An off-gas stream (Q = 353 ft³/min = 10 m³/min and T = 68°F = 20°C) from a soil venting project is to be treated by thermal oxidation or catalytic oxidation.

(a) Determine the size of the oxidizer (design operation temperature = 1,600°F = 871°C and design residence time = 0.5 seconds) in ft³ or m³.

(b) Determine the volume of the catalyst (design operation temperature = 900°F = 482°C and design space velocity = 36,000/hour) in ft³ or m³.

7.5. Soil venting is considered as another alternative to treat the excavated soil from a tank removal project. Vapor-phase activated carbon adsorption, catalytic oxidation, and thermal oxidation are considered for off-gas treatment. For the purpose of comparison, the design influent flow to the air treatment system is selected to be 100 standard ft³/min (scfm). Based on the results from air dispersion modeling coupled with risk assessment, the permitted discharge is less than 0.05 lb/day of TPH.

Determine or estimate the following:

(a) Maximum allowable TPH mass loading rate (in lb/day) to the off-gas treatment system

(b) Minimum time required to clean up the soil (in days)

Carbon Adsorption

(c) Amount of activated carbon in one carbon unit (lb)

(d) Rate of carbon spent (lb/day)

(e) Minimum number of carbon units needed

(f) Number of units and flow scheme of the carbon units (in series, in parallel, etc.)

(g) Frequency of the carbon change-out (in days)

(h) Total amount of the carbon spent for the project (in lb)

Catalytic Oxidizer

(i) Size of the catalyst

Thermal Oxidizer

(j) Size of the oxidation chamber

Use the following simplified assumptions in your calculation:

- Volume of excavated soil pile = 500 yd³
- Mass of TPH to be removed = 337.5 lb
- Automatic air dilution to keep the mass TPH loading to the treatment unit constant throughout the project (an impossible case)
- Carbon units are available in 55-gallon drums (1.8 ft in diameter and 2 ft in height)
- Activated carbon design adsorption capacity = 10 lb TPH/100 lb carbon
- Design air flow velocity through the carbon units = 60 ft/min
- Bulk density of activated carbon (30 lb/ft³)
- No make-up air is required for the incinerator
- TPH removal efficiencies of three off-gas treatment systems are the same at 98%

- Design space velocity for the catalytic incinerator = 20,000/hour
- Design combustion temperature for the incinerator = 1,600°F
- Design air residence time for the incinerator = 0.75 seconds for 98% destruction at 1,600°F

REFERENCES

Lee, C. C. and Huffmann, G. L. (2007). "Incineration Technologies and Facility Requirements." *Handbook of Environmental Engineering Calculations*, C. C. Lee and S. D. Lin, eds., McGraw-Hill Education, New York.

U.S. EPA. (1972). *Afterburner System Study*. EPA R2-72-062, NTIS PB-212-560. U.S. Environmental Protection Agency, Washington, DC.

U.S. EPA. (1991). *Handbook - Control Technologies for Hazardous Air Pollutants*. EPA/625/6-91/014. U.S. Environmental Protection Agency Office of Research and Development, Washington, DC.

U.S. EPA. (1992). *Control of Air Emissions from Superfund Sites*. EPA/625/R-92/012. U.S. Environmental Protection Agency Office of Research and Development, Washington, DC.

U.S. EPA. (2006). *Off-Gas Treatment Technologies for Soil Vapor Extraction Systems: State of the Practice*. EPA-542-R-05-028. U.S. Environmental Protection Agency Office of Superfund Remediation and Technology Innovation, Washington, DC.

8 Long-Term Monitoring, Operation, and Maintenance of Remediation Systems

8.1 INTRODUCTION

Chapters 2, 3, and 4 have detailed the background, risk, and investigation of hazardous waste sites, and Chapters 5, 6, and 7, the design of remedial technologies. But once a remediation system is designed and built, work should not stop. The system must be monitored, operated, and maintained from several months to even a few decades. In fact, the lifetime costs associated with monitoring, operating, and maintaining these systems are, most of the time, higher than the initial design and construction costs. Often, an engineer who works on all aspects of remediation projects spends most of his or her career in operation and maintenance (O&M) rather than design. The problems in this chapter focus on analyzing data collected in the field to evaluate the performance of remediation systems.

It is important to understand the definitions of monitoring, operation, and maintenance, and how some of those functions overlap. *Monitoring* refers to conducting sampling and analyzing data on chemicals of concern (COCs) to evaluate whether the remediation goals are being met. *Operation* refers to the management of the remediation system parameters, such as flow rate, pressure, and temperature, so that it runs efficiently. *Maintenance* refers to routine upkeep of equipment, like lubricating pump parts. These three activities involve routine and non-routine tasks. Routine tasks are carried out in consistent intervals. For example, groundwater monitoring usually occurs quarterly (every three months), and regular O&M occurs weekly. Non-routine tasks would be additional sampling outside of regular intervals to identify reasons for unpredictable COC decrease, evaluation of remediation strategy replacement to optimize the system, and replacement of major equipment.

There is a natural overlap of monitoring, operation, and maintenance activities, and one does not usually need to be concerned over which word describes the activity. Ultimately, the goal is to perform these activities effectively so that the remediation systems meet their goals in the long term.

This chapter is organized in the order of some of the remediation techniques covered in Chapters 5 and 6 and frequently refers to concepts introduced in them. Some of the monitoring and O&M methods can be applied to remediation techniques not mentioned in this chapter. However, those covered here are representative of many types of systems.

8.2 SOIL VAPOR EXTRACTION

Monitoring and O&M of soil vapor extraction (SVE) systems focus on analyzing the effectiveness of the system in removing COC mass from the subsurface by recording common operational parameters, such as flow rates, temperature, and pressure, and maintaining the equipment so that it runs smoothly. The subsequent sections provide a few examples.

8.2.1 MONITORING THE REMOVAL OF COC MASS THROUGH SVE

Part of the monitoring of a remediation system is to examine its effectiveness in removing a contaminant from the subsurface. For SVE systems, that means measuring COCs extracted from the soil before the off-gas is treated.

The field procedure for collecting the necessary data includes the following items:

- Collect a vapor sample from each extraction well or the combined flow of the sum of all extraction wells. This can be done at the manifold (Figure 8.1a shows a piping manifold, and Figure 8.1b shows a sample port). A sample port on the manifold is opened and connected to a portable pump via a tube. The pump extracts the vapor running through the pipe and fills a specialized thermoplastic bag (e.g., Tedlar® bag). The sample is then shipped or delivered to a certified environmental analytical laboratory.
- Measure and record the velocity or flow rate (Figure 8.2 shows a velocity meter) from the location(s) on the pipe where the sample was collected. If

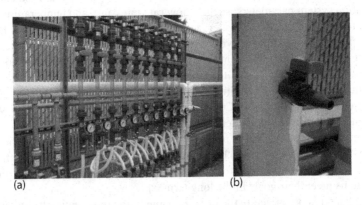

(a)   (b)

FIGURE 8.1 (a) An SVE piping manifold. All vertical pipes shown are piped from vapor extraction wells. The horizontal pipe joining the vertical pipes carries the combined flow from all the wells. (b) Each pipe has a port from which to collect a vapor sample. (Photo credit: Michael Shiang.)

FIGURE 8.2 The pictured instrument is a velocity meter. Velocity is measured inside the pipe when the metal probe is inserted into a port opening in the pipe. The digital screen displays the velocity and, in some models, temperature and flow rate. (Photo credit: Michael Shiang.)

the field instrument measures only the velocity and not the flow rate, then the flow rate can be calculated as

$$Q_a = v_a A \tag{8.1}$$

where Q_a is the actual (measured) flow rate, v_a is the measured velocity, and A is the inside cross-sectional area of the pipe.

- Measure and record the temperature and pressure from the location(s) on the pipe where the sample was collected. The pressure reading is negative if the sampling point is in a vacuum. Gauges and field instruments are used in making these measurements.

Once the data are collected, the following steps are used to calculate the mass removed from the soil via SVE. This is a standard data analysis procedure when operating an SVE system. Section 7.3.1 also has a relevant discussion on this topic.

Step 1. Take the pressure reading from the gauge (P_g) and convert it to absolute pressure (P_{abs}) using the following equation:

$$P_{abs} = P_g + P_{atm} \tag{8.2}$$

where P_{atm} is atmospheric pressure (1 atm = 14.7 psi = 29.9 in-Hg = 760 mm-Hg = 101.3 kPa).

Step 2. Calculate the standard flow rate, Q_s, which is the actual flow rate, Q_a, corrected to standard temperature, T_s, and pressure, P_s, as shown in Eq. 7.8 in Chapter 7:

$$Q_s = Q_a \left(\frac{P_a}{P_s} \times \frac{T_s}{T_a} \right) \qquad (7.8)$$

P_a and T_a are the measured pressure and temperature taken at the field. Making this correction becomes the bulk of the calculations.

Step 3. If the laboratory results for the chemicals in the vapor are reported in units of ppmV, convert them to mass/unit volume of air, typically mg/m^3 or lb/ft^3, as described in Section 2.4.3.

Step 4. Using laboratory results for each chemical detected, perform the following calculation:

$$\dot{M} = G_{COC} \times Q_s \qquad (8.3)$$

where $\dot{M}$ is the mass removal rate of the chemical, G_{COC} is the concentration of the gaseous chemical expressed in mass/unit volume of air, and Q_s is the gas flow rate adjusted to standard temperature and pressure.

To monitor whether the SVE system is operating effectively by removing COC mass from the subsurface, plotting long-term data is required. The concentration of COC in the extracted vapor should decrease as a first-order decay function, as shown in Figure 8.3. When $\dot{M}$ is calculated on a weekly or monthly basis, then the cumulative mass removed can be plotted with the expected shape of an increasing asymptotic logarithmic function, also shown in Figure 8.3. Figure 8.3 shows smooth curves, but the plots are built with points on the graph. The data are usually noisy, with the

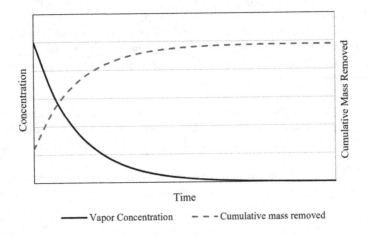

Time

———— Vapor Concentration   – – – Cumulative mass removed

FIGURE 8.3 The expected behavior of chemical concentrations and mass removal at an SVE site.

points fluctuating quite a bit. But the long-term best-fit curves should look similar to Figure 8.3.

Example 8.1: Estimating Mass of COC Removed from the Soil Using SVE

A vapor sample was collected from a pipe with extracted soil vapor. The measured flow rate was 50 cubic feet per minute (cfm) (1.4 m³/min), the temperature was 70°F (21°C), and the gauge pressure was −15 in-Hg (-381 mm-Hg) of vacuum. The COC was acetone, at a concentration of 14,000 ppmV. Acetone has a molecular weight of 58.08 g/mol. If these conditions are stable for one week, what is the total mass (lb and kg) of acetone removed from the subsurface?
Use standard temperature as 77°F (25°C) and standard pressure as 29.9 in-Hg (=14.7 psi = 760 mm-Hg = 1 atm).

SOLUTION:

(a) Convert the gauge pressure reading to absolute pressure using Eq. 8.2:

$$P_{abs} = -15\,\text{in-Hg} + 29.9\,\text{in-Hg} = 14.9\,\text{in-Hg}$$

$$P_{abs} = -381\,\text{mm-Hg} + 760\ \text{mm-Hg} = 379\ \text{mm-Hg}$$

(b) Next, find Q_s using Eq. 7.11. Note that any units of pressure can be used. Rankine units of temperature must be used for USCS, and Kelvin for SI.

$$Q_s = 50\frac{\text{ft}^3}{\text{min}} \times \frac{14.9\,\text{in-Hg}}{29.9\,\text{in-Hg}} \times \frac{(460+77)R}{(460+70)R} = 25\,\text{standard}\frac{\text{ft}^3}{s} = 25\,\text{scfm}$$

$$Q_s = 1.4\frac{m^3}{\text{min}} \times \frac{379\ \text{mm-Hg}}{760\ \text{mm-Hg}} \times \frac{(273+25)K}{(273+21)K} = 0.71\,\text{standard}\frac{m^3}{s}$$

(c) Convert the acetone concentration from ppmV to lb/ft³ (mg/m³). Once the vapor sample is collected, it is assumed that it is at standard temperature and pressure. Use Eq. 2.6 (and 2.5 for SI):

$$G\left(in\frac{\text{lb}}{\text{ft}^3}\right) = G(\text{inppmV})\left(\frac{MW}{MV(\text{inft}^3)}\right) \times 10^{-6}$$

$$G\left(in\frac{\text{mg}}{m^3}\right) = G(\text{inppmV})\left(\frac{MW}{MV(\text{inL})}\right)$$

First, the molar volume of air can be found using the Ideal Gas Law (Eq. 2.7b):

$$MV = \frac{RT}{P} = \frac{\left(10.731\dfrac{\text{ft}^3 \cdot \text{psi}}{\text{lb}-\text{mole}\cdot R}\right)(460+77)R}{14.7\,\text{psi}} = 392\,\text{ft}^3/(\text{lb}-\text{mole})$$

$$MV = \frac{RT}{P} = \frac{\left(0.08206\frac{L\cdot atm}{mol\cdot K}\right)(273+25)K}{1atm} = 24.45\,L/mol$$

Now use Eq. 2.6 for USCS units and Eq. 2.5 for SI units. Remember that conversion factors are already embedded in these equations:

$$G\left(in\frac{lb}{ft^3}\right) = 14,000\,ppmV\left(\frac{58.08\,lb/lb-mol}{392\,ft^3/lb-mol}\right)\times10^{-6} = 0.0021\,lb/ft^3$$

$$G\left(in\frac{mg}{m^3}\right) = 14,000\,ppmV\left(\frac{58.08\,g/mol}{24.45\,L/mol}\right) = 3.32\times10^4\,mg/m^3$$

(d) Finally, calculate the mass flow rate and the mass removed in 7 days:

$$\dot{M} = 0.0021\frac{lb}{ft^3}\times25\frac{sft^3}{min} = 0.052\frac{lb}{min}$$

$$\dot{M} = 3.32\times10^4\frac{mg}{m^3}\times0.71\frac{m^3}{min} = 2.35\frac{mg}{min} = 0.024\frac{kg}{min}$$

Convert lb/min to lb/week (kg/min to kg/week):

$$\dot{M} = 0.052\frac{lb}{min}\times\frac{60\,min}{1h}\times\frac{24h}{1day}\times\frac{7\,days}{1week} = 530\frac{lb}{week}$$

$$\dot{M} = 0.024\frac{kg}{min}\times\frac{60\,min}{1h}\times\frac{24h}{1day}\times\frac{7\,days}{1week} = 240\frac{kg}{week}$$

Discussion: Because of the relatively high vacuum (-15 in-Hg and −381 mm-Hg), the standard flow rate magnitude is about half of the measured flow rate.

The types of calculations described in the previous example can be done frequently to identify the asymptotic behavior shown in Figure 8.3. When data are collected for a long enough period of time that an asymptote is observed, a typical follow-up is to optimize the system, which may include replacing pumps and other equipment, evaluating whether some low-concentration extraction wells can be shut off, consider pulsed operation, or consider using another treatment system entirely. A pulsed operation involves the shut-down of the system for weeks or months, followed by a restart. Usually, a rebound effect is observed upon restarting, meaning that vapor COC concentrations increase after the down-time, though not to the high levels found initially. This condition warrants running the SVE system in a pulsed manner (a few months on and a few weeks off, for example) until no rebound effect is detected.

8.2.2 Cost Considerations of SVE Operations

In the previous section, we saw that the efficiency of SVE operations decreases when the cumulative mass removal becomes asymptotic, requiring optimization to make the system more efficient. One way to consider optimization is to consider the annual costs of operation. More aspects of remediation process optimization (RPO) are described in Chapter 9.

Operational costs include costs for labor, electricity, supplies, and sample analysis. It is wise to consider annual operation costs against the mass of COC removed. We see from Figure 8.3 that the largest mass removal occurs at the beginning of the remediation activity. But if the cost of running the SVE system is about the same every year, then the cost per unit mass removed is lowest in the first years of operation. The next example demonstrates this point.

Example 8.2: Cost analysis and the need to re-evaluate the remediation system

An SVE system has run for five years. The table below shows the annual costs and COC mass removed. The cost of the first year includes the construction of the system. (a) Calculate the cost of removal for each year. (b) In hindsight, when should a remediation professional have considered optimizing the system? Explain your answer.

Year	Remediation Cost ($)	Mass Removed (kg)
1	500,000	100
2	150,000	50
3	150,000	10
4	150,000	5
5	150,000	1

SOLUTION:

(a) Take the annual cost and divide it by the COC mass removed:

Year	Remediation Cost ($)	Mass Removed (kg)	$/kg
1	500,000	100	5,000
2	150,000	50	3,000
3	150,000	10	15,000
4	150,000	5	30,000
5	150,000	1	150,000

(b) Remediation optimization should happen around the 3rd year because the $/kg cost has increased by one order of magnitude.

Discussion: Deciding when to start optimization at a real site should depend on more measures than just costs. For example, if a system turns out not to be remediating parts of the plume, the system must be optimized. This example shows just one way to evaluate system efficiency. The treatment system should certainly be reconsidered by the 5th year when it costs $150,000 to remove just 1 kg of COC!

8.2.3 OPERATION AND MAINTENANCE CHECKS FOR SVE

One way to understand the complexity of operating a remediation system is to examine a piping and instrumentation diagram (P&ID). A P&ID shows the process controls, piping, equipment, and associated instrumentation. It is different from a site plan because it is not a construction drawing showing where the equipment is spatially located. Instead, it shows how the equipment is connected and the direction of fluid flows. The P&ID in Figure 8.4 shows an SVE system and air sparging system installed at a site. In this section, we focus on the SVE components.

Following the P&ID according to the path of the soil vapor, and carefully reading the labels and legend, we start towards the bottom right of the drawing and note the three "Slotted Vertical Extraction Vent Pipes" (or vapor extraction wells). Vapor from the soil formation flows into the wells, as indicated by the arrows. This vapor flow is caused by the vacuum provided by the blower, which we see later in this description. Notice that each extraction well has a pressure indicator, flow control valve, sample port, and flow meter. Those are physically located at the top of the well

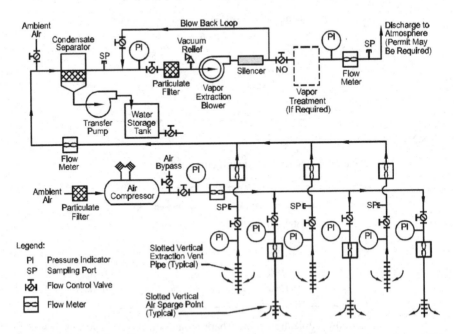

FIGURE 8.4 A piping and instrumentation diagram (P&ID) for soil vapor extraction (SVE) and air sparging. (U.S. EPA 2017.)

or in the pipe manifold so a person can easily turn the valve, read the pressure gauge and flow meter, and collect a sample.

The flow control valve can be fully open, partially open, or closed, depending on how important it is to treat that part of the site or whether flow needs to be controlled to the "vapor extraction blower" (or vacuum pump) or "vapor treatment" unit. A pressure indicator is a gauge that should indicate a vacuum (negative gauge pressure) upstream of the blower and a positive pressure downstream of the blower. Since the vapor in the pipe should be in a vacuum, an increase in pressure (more positive gauge pressure) would indicate a leak in the pipe, whereas an increased vacuum (decreased negative pressure) would indicate a clog in the well or pipe. The flow meter reads the flow rate of soil vapor moving through the wells, and the sample ports are where a person can draw out a sample of vapor to analyze for COCs. The pressure, flow, and COC concentrations are those measurements needed to calculate how much COC is being removed from the subsurface, the calculation covered in Section 8.2.1 Example 8.1. Although not shown in the P&ID in Figure 8.4, a temperature gauge can be included.

Continuing to follow the vapor flow, we see that the three extraction wells connect to a horizontal pipe that carries all of the vapor flow to the extraction and treatment system. A flow meter measures the combined flow from all wells. The next important item is the "condensate separator," also introduced as the knockout tank in Section 5.3.2. This tank separates the entrained moisture (condensate) from the vapor. The condensate falls to the bottom of the tank, while drier vapor flows out of the top of the tank. The condensate is pumped out with a transfer pump and stored in a tank. That water will later be treated or disposed of according to the concentration of COCs.

Continuing to follow the vapor flow in the P&ID, the vapor coming out of the top of the condensate tank passes through the "vapor extraction blower" (vacuum pump), which provides the vacuum and pressure to extract and move the vapor through this entire system. After moving through the blower, the vapor passes through a treatment unit, which can be any of the treatment units discussed in Chapter 7. The treated vapor is then discharged into the atmosphere. Notice that there are pressure gauges, sampling ports, and valves along the pipe. These serve as additional measurement points that help an engineer and operator determine whether the system is operating correctly, efficiently, and meeting the remediation goals.

8.2.4 RECORDKEEPING

Recordkeeping throughout the years of operation of a remediation system is extremely important.

Section 8.2.1 discusses a specific procedure for collecting and analyzing data for the effectiveness of an SVE system. The mechanical system must also be working properly. Other checks that must be made and records that must be kept include (U.S. ACE 2002):

- Pressure gauge readings throughout the piping system, at the vapor extraction wells, pipe manifold, upstream and downstream of pump and treatment units. Pressure readings higher than normal may indicate a clog in

FIGURE 8.5 Pressure being measured at a vapor monitoring well. (Photo credit: Michael Shiang.)

the system; pressure readings close to atmospheric pressure may indicate a leak in the piping.

- Pressure gauge readings at vapor monitoring wells. If the monitoring wells are located within the radius of influence of the extraction wells, those monitoring wells should be at a vacuum (see Figure 8.5).
- Check and record the water level in the knockout tank, which collects moisture from the extracted vapor. Dispose of or treat the water when the knockout-tank is nearly full.
- Maintain the blower by changing the oil, greasing the assembly, and other items following the manufacturers' recommendations. Keep records of dates and results of maintenance activities.
- Clean and replace the particulate filters. Record the dates and the conditions in which the filters were cleaned and replaced.

8.3   *IN SITU* CHEMICAL OXIDATION

Long-term monitoring and O&M of *in situ* chemical oxidation (ISCO) involve procedures to ensure safety on-site, sample soil and groundwater to detect decreased COC concentrations, keep records of chemical injection pressures and flow rates, and maintain the equipment.

Although oxidants are used in ISCO to reduce the toxicity of COCs, oxidants themselves are safety hazards during the operation of ISCO systems. For example,

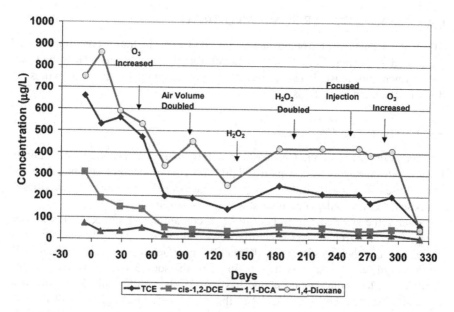

FIGURE 8.6 Monitoring data from a groundwater monitoring well for an ISCO pilot test at a Superfund site in California. (Source: URS Group 2006.)

dust and particulates from the oxidants (e.g., permanganate and persulfate) must be controlled to avoid inhalation and damage to the respiratory tract. Hydrogen peroxide can react with flammable COCs and cause fires and explosions. Therefore, the safe handling of oxidants is of utmost importance. Product labeling usually warns of such dangers. Workers should be experienced with these oxidants and wear personal protective equipment (PPE).

To evaluate the effectiveness of ISCO remediation systems, personnel collect soil and groundwater samples over a period of time. Plotting the measured data would result in a graph similar to Figure 8.6. This figure shows the COC results in groundwater in a monitoring well near an ISCO injection point where ozone and hydrogen peroxide were alternately injected. A sustained lowering of COC concentrations would indicate the success of the ISCO remediation system. If several monitoring wells are present, then the radius of influence of the chemical application through injection wells can be estimated.

Maintenance activities for the proper operation of ISCO sites include (U.S. EPA et al. 2006):

- Maintaining the oxidant-mixing apparatus and storage tanks;
- Checking piping for leaks;
- Checking the power supply for proper performance;
- Maintaining the oxidant pump; and
- Monitoring the oxidant injection pressure.

8.4   GROUNDWATER PUMP-AND-TREAT

The primary purpose of groundwater pump-and-treat is hydraulic containment, and the secondary purpose is COC mass removal from the groundwater. The groundwater treatment unit must also be monitored for proper operation. These activities are described in the next sections.

8.4.1   HYDRAULIC CONTAINMENT

As seen in the discussion in Chapter 6 on well hydraulics and capture zones, pumping groundwater alters its flow path and prevents contamination from spreading and reaching sensitive locations. Here we provide an example of monitoring activities to check whether hydraulic containment is being achieved.

Observe the site depicted in Figure 8.7. A contamination source area is located in the west, and groundwater generally flows from west to east. The contaminated groundwater is shown as the shaded area. A subsurface barrier wall has been built into the aquifer to block groundwater flow past the source area. However, this has been done after some contamination has moved east. As a result, four extraction wells are installed: EW-1 and EW-2 in the source area and upgradient of the barrier wall, and EW-3 and EW-4 near the front of the plume to the east. Fourteen monitoring wells are also installed, and their groundwater elevations are noted, in units of feet. EW-1 and EW-2 pump groundwater from the source area to maintain the groundwater elevations from rising, given that groundwater flow is blocked by the barrier wall. EW-3 and EW-4 provide hydraulic containment, so the plume does not move farther. In this example, one way to determine if hydraulic containment is working is to note that the groundwater contour line at 113.5 ft curves towards the pumping wells.

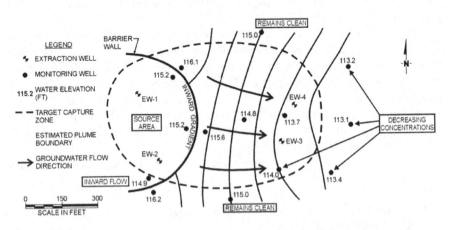

FIGURE 8.7   Lines of evidence for evaluating groundwater capture. (Source: U.S. EPA 2002.)

To ensure that the system continues to meet long-term remediation goals, person-nel must carry out many activities:

- Periodic monitoring of groundwater elevations in monitoring wells;
- Ensuring that pumps are operating properly;
- Replacing worn-out pumps/parts;
- Fixing leaks in the piping network; and
- Adding additional extraction wells if regular monitoring and O&M show that hydraulic containment is not fully achieved.

8.4.2 REMOVAL OF COC MASS THROUGH GROUNDWATER PUMP-AND-TREAT

As seen in Section 8.2.1, the mass of COCs removed from the subsurface can be determined for systems that extract a fluid from the subsurface. Groundwater pump-and-treat is one such system, and COC concentrations in the water before treatment can be used to estimate the mass removed from the aquifer.

The field procedure for collecting the data is similar to that of SVE but simpler. It includes the following items:

- Collect a groundwater sample from each extraction well or the combined flow of the sum of all extraction wells. The sample is then shipped or deliv-ered to a certified environmental analytical laboratory.
- Measure and record the velocity or flow rate from the location(s) on the pipe where the sample was collected.

Once the data are collected, a simple calculation is used to determine the mass removed from the groundwater via pumping. This is a standard data analysis proce-dure when operating a pump-and-treat system. The mass removal rate is:

$$\dot{M} = C_{COC} \times Q \tag{8.4}$$

where C_{COC} is the concentration of COC in the extracted groundwater, and Q is the water flow rate. The next example shows how this calculation can be done over mul-tiple months.

Example 8.3: Estimating Mass of COCs Removed from the Groundwater Using Pump-and-Treat over Multiple Months

You are a remediation engineer working at a site where benzene and toluene are the primary pollutants being removed from the subsurface using a pump-and-treat system. Two groundwater extraction wells are used to extract contaminated groundwater. The responsible party is pressured by the state environmental regu-latory agency to report how well the system has performed in the past 3 months in removing the contaminants from the subsurface. You have collected the fol-lowing data. What is the mass (in pounds) of benzene and toluene that have been removed from the aquifer?

	Well 1			Well 2		
Month	Benzene Conc. (µg/L)	Toluene Conc. (µg/L)	Flow rate (gpm)	Benzene Conc. (µg/L)	Toluene Conc. (µg/L)	Flow rate (gpm)
January	555	431	38	95	58	41
February	376	350	36	88	36	43
March	98	75	37	41	12	39

SOLUTION:

Apply Eq. 8.4 to each well, COC, and month, using the appropriate conversion factors, and assuming the concentration and flow rate are constant for each month. Starting with benzene and January:

$$\dot{M} = 555\frac{\mu g}{L} \times 38\frac{gal}{min} \times \frac{3.785\,L}{gal} \times \frac{lb}{453.6 \times 10^6\,\mu g} \times \frac{1,440\,min}{1\,day} \times \frac{31\,days}{1\,month}$$

$$= 7.8\frac{lbs}{month(January)}$$

This calculation is repeated for both chemicals, two wells, and three months (paying attention to use 28 days for February). Because of the repeated calculations, using a spreadsheet is recommended. The final results are the sum of all benzene and toluene masses in the table below.

	Well 1					Well 2				
Month	Benzene Conc. (µg/L)	Toluene Conc. (µg/L)	Flow rate (gpm)	Benzene Mass (lb)	Toluene Mass (lb)	Benzene Conc. (µg/L)	Toluene Conc. (µg/L)	Flow rate (gpm)	Benzene Mass (lb)	Toluene Mass (lb)
January	555	431	38	7.8	6.1	95	58	41	1.4	0.9
February	376	350	36	4.5	4.2	88	36	43	1.3	0.5
March	98	75	37	1.3	1.0	41	12	39	0.6	0.2

The total masses removed are 17.1 lbs of benzene and 12.9 lbs of toluene.

DISCUSSION:

1. The concentrations have decreased from one month to another. This behavior is typical as also seen in the SVE section.
2. Flow rates are similar for all three months and should be similar unless there are operational problems, such as well fouling or pump failures, or weather variations, such as extreme rainfall or drought.
3. The solution to this problem assumes that the concentrations and flow rates remain constant for each month and that there is a step difference in those values from one month to another.

8.4.3 MONITORING FOR ACTIVATED CARBON EXHAUSTION

As seen in Chapter 6, granular activated carbon (GAC) is a common groundwater treatment technology. To determine when a GAC vessel needs to be replaced, there must be periodic sampling of the influent and effluent ports. The influent sample helps us calculate the mass of COCs removed from the groundwater, as indicated in Section 8.4.2. The effluent sample helps us understand the mass of COCs retained inside the carbon vessel. We will focus on the carbon vessel in this section.

Let's picture the behavior of one carbon vessel over time, given a volume of water running through it, as shown in Figure 8.8. The horizontal axis represents the volume of treated water passing through the vessel. It can also represent the passing of time. The vertical axis represents the ratio of the COC concentrations in the effluent and influent ports of the vessel. When C_{out}/C_{in} is zero, no COCs are in the effluent samples. When C_{out}/C_{in} is 1.0, the effluent concentration is equal to the influent concentration, and the GAC is said to be exhausted. A GAC vessel should only be operated until this point if a secondary vessel is connected to it in series.

As water flows through the vessel from top to bottom, the COC adsorbs to the GAC near the top of the vessel. With time, the COC will continually adsorb to GAC lower and lower into the vessel. In Figure 8.8, this is shown in the vessel drawn above V_1 in the horizontal axis. The color gradient, from dark to light, at the top of the vessel represents the adsorption zone, or mass transfer zone (MTZ), where the COC is adhering to the GAC. The lower part of the GAC vessel is drawn with a light color indicating that COCs have not reached that part of the vessel yet.

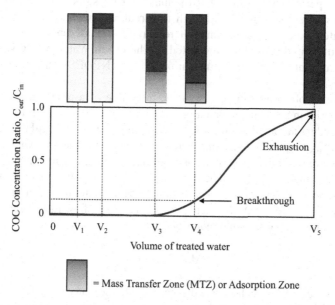

FIGURE 8.8 Demonstration of the mass transfer zone (MTZ) in a carbon vessel.

At V_2, we see that the top of the GAC is fully saturated, as indicated by the dark color, and the MTZ has moved downward. The bottom of the vessel is still clear of COCs, so the effluent concentration is still zero.

At V_3 the MTZ has moved to the bottom of the vessel, and we see a small concentration of COC in the effluent sample. At V_4, the MTZ has continued to move down, and a higher concentration of COC leaves the effluent. We labeled this point "breakthrough" to signify a concentration limit. That is, if this vessel were the final vessel before the treated water is discharged to a stream, the GAC should be changed out before the breakthrough concentration.

If we were to leave the vessel operating, at the point represented by V_5, the effluent COC would be nearly equal to the influent COC. Remember that in Chapter 6 we mentioned that GAC vessels are installed with at least two in series (see Figure 6.9). This breakthrough scenario is exactly why another vessel, or "secondary" vessel, is necessary. This way, the primary vessel is fully and efficiently used, and the secondary vessel prevents COCs from being discharged from the site. At this point, the system is shut down temporarily so that the exhausted GAC can be replaced with new GAC, and the piping configuration and valves are changed so that the secondary vessel now becomes the primary vessel, and vice-versa.

8.4.4 MONITORING FOR DISCHARGE INTO A RECEIVING WATER BODY

So far, Section 8.4 has discussed hydraulic containment, removal of COC mass from groundwater, and monitoring for the exhaustion of GAC. Another important operational parameter is the mass of COCs discharged into a surface water body after treatment. Although the treatment technologies for cleaning groundwater are effective, at times there may be a small amount of COC being discharged. This discharge is regulated by the Clean Water Act (CWA) and its National Pollutant Discharge Elimination System (NPDES) permit, as described in Section 3.4.3.

Therefore, collecting samples from the discharge of a treatment system, be it GAC, air stripping, metal precipitation, or advanced oxidation, is a routine operational procedure at remediation sites. The sampling results are regularly reported to the permitting agency (usually the State's environmental agency), which regulates the flow rate of water discharge and concentration, thus the loading rate of the COCs, into the receiving water body. The flow rate is measured using a gauge in the field. Concentrations are determined by collecting a sample and having it analyzed in a laboratory. The loading rate is calculated just as a mass flow rate is calculated, for example, using Eq. 8.4 shown previously:

$$\dot{M} = C_{COC} \times Q \tag{8.4}$$

where $\dot{M}$ is the COC mass flow rate discharged into the water body, C_{COC} is the COC concentration after treatment (usually a number below the discharge limit), and Q is the flow rate leaving the treatment system. Routine measurements will yield the types of data much like those shown in Example 8.3 but with much smaller concentrations. Long-term data will indicate how much mass of COC the water body is receiving during the life of the remediation project.

8.5 AIR SPARGING

Air sparging is an *in situ* technology in which ambient air is injected into the groundwater to volatilize and aerate COCs. The COCs can be biodegraded *in situ* or volatilized into the vadose zone. Because COCs rise to the vadose zone, air sparging is most often applied together with soil vapor extraction (SVE).

To visualize how air sparging is combined with SVE and what operational parameters need to be monitored, refer to Figure 8.4, a P&ID that depicts both technologies. The SVE portion of the P&ID was discussed in Section 8.2.3.

Following the P&ID according to the path of the injected air and carefully reading the labels and legend, we start towards the center-left of the drawing and note the words "Ambient Air." This is atmospheric air that is suctioned through a particulate filter to remove dust and then stored in an air compressor tank. The air compressor is a piece of equipment that compresses air by pressurizing it into a tank. That pressurized air then leaves the tank into a pipe that is connected to three "Slotted Vertical Air Sparge Points" (or air sparging wells or injection wells). Each injection well has a flow control valve, pressure indicator, and sampling port. These appurtenances are used to control the air flow and measure how effectively air is being injected into the aquifer. The arrows out of the bottom of the injection wells show that the air enters the aquifer.

As an example of what operational parameters should be checked for system effectiveness, Figure 8.9 shows a record sheet that can be filled out by technicians or engineers when inspecting air sparging sites. The wellhead pressure and air flow readings on the bottom of the list are important parameters that may indicate clogging in the wells. Many groundwater aquifers naturally contain reduced iron (Fe^{2+}). But when the ferrous ion is exposed to oxygen from the injected air, it is oxidized into insoluble ferric iron (Fe^{3+}) oxide. Fe^{3+} can form a precipitate and cause clogging of pores within the soil and in the screen and filter of the air sparging well. Therefore, air sparging is most effective when the concentration of Fe^{2+} in the groundwater is less than 20 mg/L. If clogging in the well is detected, it may need to be replaced with a new well (U.S. EPA 2017).

For long-term monitoring to determine whether COC concentrations in groundwater are decreasing, quarterly (every three months) sampling of nearby groundwater monitoring wells is necessary. This should not be a one-time event, as COC concentrations might be low when air is being injected, but might rebound when the system is shut down. Monitoring for air sparging effectiveness is very similar to monitoring for ISCO effectiveness, as described in Section 8.3 and Figure 8.6. Effectiveness is demonstrated by detecting sustained low COC concentrations in surrounding monitoring wells.

Example IAS System Operational Checklist Mechanical System Measurements

Inspector name:			Date:	

Item	Time Checked	Typical Values*	Initial Reading	Reading After Any Adjustments
Compressor/Blower Discharge Pressure		8 psi		
Compressor/Blower Discharge Flow @ Pressure Above		100 cfm		
Sparge Blower Discharge Temp.		240°F		
Bearing Oil Temperature		200°F		
Bearing Oil Pressure		20 psi		
Interval Operating Hours		—		—
Motor Amps		8		
Oil Level		—		
Aftercooler Inlet Pressure		7 psi		
Aftercooler Inlet Temperature		180°F		
Aftercooler Outlet Pressure		6 psi		
Aftercooler Outlet Temperature		120°F		
Ambient Air Temperature (outside/inside shed)		—		—
IAS-1[1] Wellhead Pressure		5.5 psi		
IAS-1[1] Wellhead Air Flow		6 cfm		
IAS-2[1] Wellhead Pressure		7.4 psi		
IAS-2[1] Wellhead Air Flow		2 cfm		

Notes: 1. Each other IAS well should be listed individually.
       2. Operator should operate valves and controls at least once each month.

* Values shown for example only, column to be filled in according to actual typical measurements

FIGURE 8.9   Sample checklist for proper *in situ* air sparging (IAS) operation. (Source: U.S. ACE 2013.)

8.6   MONITORED NATURAL ATTENUATION

Monitored natural attenuation (MNA) is an active remediation system only because soil and groundwater sampling are used to detect whether COC concentrations decrease by natural processes, such as dispersion, dilution, sorption, volatilization, biodegradation, and chemical reactions. As such, there is no equipment to be operated and maintained.

The main lines of evidence for MNA are (NJDEP 2012):

- Direct lines of evidence: COC plume is stable or shrinking; plotting COC trends decreasing over distance and time.
- Indirect lines of evidence: Geochemical analysis indicates favorable conditions for natural attenuation processes.

For checking a direct line of evidence, the following procedure can be followed (NJDEP 2012; Wisconsin DNR 2014):

Step 1. Estimate the groundwater seepage velocity and COC velocity using Darcy's law and the retardation factor, as described in Chapter 4.

Step 2. Measure the distance between the monitoring well closest to the edge (but outside of) the plume and the nearest contaminated well along the groundwater flow path.

Step 3. From the information in Steps 1 and 2, calculate the time frame in which the COC movement is likely to be detected in the downstream well.

Step 4. Monitor COC concentrations and geochemical parameters for a time greater than the time period calculated in Step 3.

Step 5. Plot the data in graphs to determine whether the COC concentrations are decreasing at an acceptable rate. A typical concentration decrease curve is the first-order decay curve shown by

$$C_t = C_o e^{-kt} \tag{8.5}$$

where C_t is the COC concentration at time t, C_o is the initial concentration, k is the decay rate constant, and t is time.

If COC levels are decreasing, then the natural processes of natural attenuation are occurring. If the COC levels are not decreasing, then other more active remedial measures are necessary to control the COC.

Example 8.4: Estimate the extent of natural attenuation using data plots

A monitoring well in a benzene plume has the concentrations shown below. (a) Determine the first-order decay rate constant. (b) Estimate how long, from the initial monitoring date, it will take for the benzene concentration to decrease to 5 µg/L.

Date	Days Elapsed	Benzene (µg/L)
1/5/2018	0	4,900
3/5/2018	59	3,700
6/5/2018	151	400
9/5/2018	243	180
12/5/2018	334	140
3/5/2019	424	50
6/6/2019	517	90
9/2/2019	605	60
12/8/2019	702	40
2/28/2020	784	10

SOLUTION:

This problem follows Steps 4 and 5 described above. This problem eliminates Steps 1 and 2 because, instead of looking at a COC decay over distance, it examines a COC decay over time.

(a) Use Eq. 8.5 to determine the decay rate:

$$C_t = C_o e^{-kt}$$

$$10 = 4,900e^{-k(784)}$$

$$0.00204 = e^{-k(784)}$$

$$\ln(0.00204) = -784k$$

$$k = 0.0079\,\mathrm{day}^{-1}$$

(b) Use Eq. 8.5 again, this time to determine the time it would take to achieve the desired concentration of 5 μg/L:

$$5 = 4900e^{-0.0079t}$$

$$0.00102 = e^{-0.0079t}$$

$$t = 872\,\mathrm{days}$$

DISCUSSION:

1. The concentrations have decreased each month, which indicates that natural attenuation is occurring.
2. The first-order decay behavior is typical for concentration decreases in remediation, also seen in the SVE section, Figure 8.3.
3. If benzene concentrations have not stabilized below 5 μg/L in 872 days, then MNA should continue past 872 days. The decay constant can be recalculated and updated over time.
4. Alternatively, this problem can be solved with graphing, as shown in Figure 8.10. This is a semi-log plot with a trendline created automatically using a spreadsheet program. The trendline equation is shown in the upper-right: 2,232 is the y-intercept of the trendline, equivalent to C_o of the trendline; and 0.007 is the calculated decay rate constant. Using $y = C_t = 5$ μg/L, solve for $x = t = 872$ days.

The previous example and paragraphs show how to use direct lines of evidence from concentration data to determine the extent of natural attenuation. This direct line of evidence does not identify which processes are causing natural attenuation. To identify if natural attenuation is caused by biodegradation or chemical reactions, an indirect line of evidence can be used. For this, geochemical parameters,

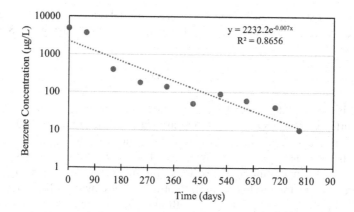

$$y = 2232.2e^{-0.007x}$$
$$R^2 = 0.8656$$

FIGURE 8.10 A semi-log plot of the benzene concentrations and associated trendline (Example 8.4).

specifically terminal electron acceptors (TEA), are monitored. Introduced briefly in Section 6.5.1, the monitoring of TEAs will be described here.

When organic compounds degrade, this happens via a reaction called oxidation-reduction reaction, or redox. This is also the type of reaction that occurs in *in situ* chemical oxidation (ISCO), introduced in Section 5.5.1. Reviewing, the organic compound is oxidized (loses electrons). As a result, another compound must be reduced (receives electrons). This reduced compound is the terminal electron acceptor (TEA). Typical TEAs are oxygen, nitrate, sulfate, and ferric iron (Fe^{3+}) minerals in the aquifer. The point of monitoring TEAs in addition to COCs is to determine if biodegradation or natural chemical oxidation is occurring. A decrease in TEAs indicates microbiological activity in the subsurface.

8.7 MONITORING FOR AIR EMISSIONS FROM OFF-GAS TREATMENT

So far, we have discussed SVE and its ability to remove the mass of COCs from the vadose zone. But another important operational parameter is the mass of COCs emitted into the atmosphere after off-gas treatment. Recall that Chapter 7 described GAC and thermal and catalytic oxidation as types of treatment of contaminated vapors. Although these technologies for treating vapors are effective, at times there may be a minor amount of COC leaving the off-gas treatment. These emissions are regulated by the Clean Air Act (CAA) or the Resource Conservation and Recovery Act (RCRA), described in Sections 3.4.5 and 3.4.6.

Therefore, collecting samples from the discharge of the off-gas treatment system, be it GAC or thermal or catalytic oxidation, is a routine operational procedure at remediation sites. The sampling results are regularly reported to the permitting agency (usually the State's environmental agency), which regulates the maximum vapor emission flow rate and concentration and mass flow rate of COCs into the air. The flow rate is measured using a gauge in the field. Concentrations are determined by collecting a sample and having it analyzed in a laboratory. The loading rate is calculated just as a mass flow rate is calculated, for example, using Eq. 8.3 shown previously:

$$\dot{M} = G_{COC} \times Q_s \qquad\qquad (8.3)$$

where $\dot{M}$ is the mass flow rate of the COC into the air and G_{COC} is the concentration of the emitted gaseous COC expressed in mass/unit volume of air, and Q_s is the gas flow rate emitted to the atmosphere, adjusted to standard temperature and pressure. Note that the actual temperature and pressure of the emitted air will be very different from the actual temperature and pressure within the treatment system. G_{COC} should be a number below the discharge limit. Routine measurements will yield data much like those shown in Example 8.1, but with much lower concentrations. Long-term data will indicate how much mass of COC enters the atmosphere during the life of the remediation project.

8.8   SUMMARY

This chapter addressed the important long-term activities that must be followed for the success of a remediation program. We addressed this by showing data that are commonly collected in the field and analyzed to determine remediation effectiveness. While not every remediation technology was addressed, the monitoring and O&M activities described apply to all remediation technologies. Major points addressed were:

- Definition of monitoring, operation, and maintenance;
- Soil vapor extraction monitoring and O&M activities, including mass removal determination and cost considerations;
- *In situ* chemical oxidation monitoring and O&M activities;
- Groundwater pump-and-treat monitoring and O&M activities, including hydraulic containment and mass removal determination;
- Air sparging monitoring and O&M activities; and
- Monitored natural attenuation (MNA) activities.

8.9   PROBLEMS AND ACTIVITIES

8.1.   Calculate the volume that 1 mole of an ideal gas occupies inside a pressurized pipe at a temperature of 55°C and an absolute pressure of 305 mm-Hg (note that since the pressure is less than atmospheric pressure, 760 mm-Hg, then the pressure is considered to be a vacuum).

8.2.   Calculate the volume that 1 mole of an ideal gas occupies inside a pipe under a temperature of 40°C and an absolute pressure of 2 atm.

8.3.   Based on your answer to problem 1, convert a concentration of benzene of 98.0 ppmV to mg/m³ and calculate the mass flow rate for a standard volumetric flow rate of 0.05 m³/s.

8.4. Based on your answer to problem 2, convert a concentration of vinyl chloride of 48 µg/m³ to ppmV.

8.5. A soil vapor extraction project uses granular activated carbon (GAC) to treat the off-gas concentrations of xylene (MW 106 g/mol) from 800 ppmV (at P=0.9 atm and T=30°C) to 100 ppmV (at STP: P=1 atm and T=25°C) at an STP flow rate of 0.09 m³/s. Calculate the mass of xylene retained in the GAC per day.

8.6. Given the graph shown (Figure 8.11) on the cumulative mass of contaminants removed using SVE, at what point should a remediation professional consider optimizing the system? Explain your answer.

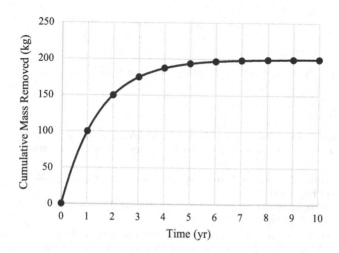

FIGURE 8.11 A graph of cumulative mass removal over time (Problem 8.6).

8.7. For the graph of COC removal over a period of 20 years (Figure 8.12), which portions of the curve indicate poor efficiency in the system? Where does the curve indicate that a remediation process optimization was performed?

8.8. Several monitoring wells located in the direction of the flow of a benzene plume have the concentrations at one point in time shown below. (a) Determine the first-order decay rate constant with respect to distance from the source. (b) Estimate at what distance from the source the toluene concentration is expected to reach 5 µg/L.

Distance along flow path (m)	Toluene (µg/L)
0	9800
28	6400
55	6000
80	500
120	750
300	12

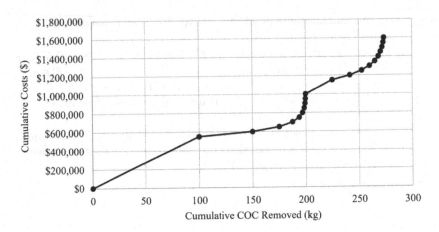

FIGURE 8.12   A graph of cumulative costs vs cumulative COC removed (Problem 8.7).

REFERENCES

NJDEP. (2012). *Site Remediation Program Technical Guidance, Monitored Natural Attenuation Technical Guidance.* New Jersey Department of Environmental Protection, Trenton, NJ.

U.S. ACE. (2002). *Soil Vapor Extraction and Bioventing. Engineer Manual.* U.S. Army Corps of Engineers.

U.S. ACE. (2013). *In-situ Air Sparging Engineer Manual, EM 200-1-19.* U.S. Army Corps of Engineers, Washington, DC.

U.S. EPA. (2002). *Elements for Effective Management of Operating Pump and Treat Systems,* EPA 542-R-02-009. U.S. Environmental Protection Agency Office of Solid Waste and Emergency Response, Cincinnati, OH.

U.S. EPA. (2017). *How To Evaluate Alternative Cleanup Technologies For Underground Storage Tank Sites, Chapter VII Air Sparging.* EPA 510-B-17-003. U.S. Environmental Protection Agency Office of Land and Emergency Management, Washington, DC.

U.S. EPA, Huling, S., and Pivetz, B. (2006). *Engineering Issue: In-Situ Chemical Oxidation.* EPA/600/R-06/072, U.S. Environmental Protection Agency Office of Research and Development, Cincinnati, OH.

URS Group, I. (2006). *Field Pilot Study of In Situ Chemical Oxidation Using Ozone and Hydrogen Peroxide to Treat Contaminated Groundwater at the Cooper Drum Company Superfund Site.* Sacramento, CA.

Wisconsin DNR. (2014). *Guidance on Natural Attenuation for Petroleum Releases. PUB-RR-614.* Wisconsin Department of Natural Resources, Madison, WI.

9 Strategies for Sustainable Remediation Projects

9.1 INTRODUCTION

So far, this book has addressed how sites become contaminated, chemical and physical aspects of contaminant movement, site investigations, risk, vadose zone remediation, groundwater remediation, off-gas treatment, and site monitoring and operation and maintenance (O&M). In this chapter, we cover some strategies for combining remediation techniques to address multiple remediation goals, plus considerations regarding cost, environmental justice, and sustainability. Therefore, this chapter provides an additional environmental, social, and economic context to the materials covered in the previous chapters.

9.2 FEASIBILITY STUDY

A feasibility study (FS) is a first step towards formulating a cleanup strategy for a contaminated site. The FS, also briefly described in Section 3.3.2, follows the remedial investigation (RI) process and development of the site conceptual model (SCM) described in Chapter 4. These combined processes are often referred to as the RI/FS process. To perform an FS, an engineer has to already know the various remediation techniques available to potentially remediate the site. Chapters 5, 6, and 7 covered those remediation techniques. This section describes in more detail how to follow the CERCLA FS process once the RI has been performed, an SCM has been developed, and potential remediation techniques are known (U.S. EPA 1988). An FS considers human health and the environment as a priority, as shown in Figure 9.1 (U.S. EPA 1990).

The FS consists of a detailed analysis of remedial alternatives. The feasibility study starts with a screening process, where many remedial technologies are assessed for adequacy for the site conditions. Sometimes the screening process starts while the RI is still in process. As the RI is refined, so is the screening of possible technologies.

After the SCM is more and more refined, and several technologies have been established as adequate for the site conditions, then it is time for a more detailed analysis of remedial alternatives. Remedial alternatives consist of a combination of technologies and management practices that fulfill the criteria shown in Figure 9.1.

The balancing, or technical, criteria are those that technical and engineering staff mostly work on. Some work items for those criteria are listed below (U.S. EPA 1990).

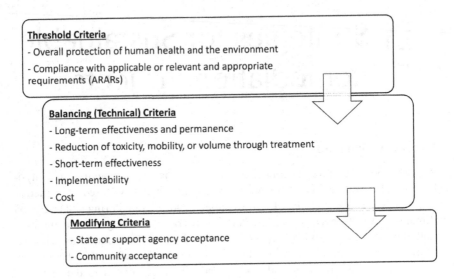

FIGURE 9.1   Criteria for detailed analysis of remedial alternatives.

Long-term effectiveness and permanence

- Does the remedial alternative reduce residual risk?
- Does the remedial alternative adequately and reliably control COCs?

Reduction of toxicity, mobility, or volume through treatment

- Does the remedial alternative treat COCs?
- Does the remedial alternative reduce the mass of COCs by destroying or treating them?
- By how much will the remedial alternative reduce the concentrations/masses of COCs?
- Is the treatment provided by the remedial alternative irreversible?
- What types and quantities or residuals remain from using the remedial alternative?

Short-term effectiveness

- Is the community protected during remediation?
- Are workers protected during remediation?
- Are there environmental impacts from the remediation?
- How long will it take for the remedial alternative to meet the objectives?

Implementability

- Is the technology able to be constructed and operated?
- How reliable is the technology?

- How easy would it be to undertake additional remedial alternatives?
- Can the effectiveness of the remedial alternative be monitored effectively?
- Will agencies approve the remedial alternative?
- Are services and materials available to build and operate the remedial alternative?

Cost

- What is the expected capital cost of the remedial alternative?
- What is the expected operation and maintenance cost of the remedial alternative?
- What is the net present worth of the remedial alternative?

These FS criteria have been in effect since the 1980s and continue to help environmental professionals, government environmental agencies, and communities strategize how to best remediate a site. With these criteria in mind, the next section describes how to select remedial alternatives.

9.3 TYPICAL REMEDIAL STRATEGIES AND COMBINING REMEDIAL TECHNIQUES

Many remediation technologies were described in Chapters 5 through 7, ranging from vadose zone remediation, saturated zone (groundwater) remediation, and off-gas treatment for technologies used for both zones. These techniques are not usually used alone and can be used in combination with one another. Two or more technologies can be combined at the same time, and sometimes one technology is used in the first few years of remediation and another is used in the later years of remediation. This section describes some strategies for selecting remedial alternatives.

As can be seen from the scenarios in the next sections, a combination of remedial technologies is necessary to address sensitive receptors and remove sources of contamination. The combination of remedial technologies to address many needs is called a *remedial strategy*. Selecting the remedial strategy should follow recommendations from the FS, described in Section 9.2, and its nine criteria, with consideration to the appropriateness of the technologies to treat the types of COCs and the type of media (e.g., soil, groundwater) affected.

9.3.1 Prevent the Contamination of Sensitive Receptors

A properly conducted site conceptual model (SCM) identifies sensitive receptors. These receptors can be water-based, such as drinking water wells, recreational lakes, and rivers used for fishing. Or they can be people-oriented, such as an elementary school exposed to vapor intrusion (the entry of underground chemical vapors through cracks in the foundation of a building).

If sensitive receptors are at imminent risk, the remedial action may be different than if sensitive receptors are months or years away from being affected (for example, a contaminated groundwater plume moving at one meter per day towards a well 1,000 m downgradient). Next, we look at a few scenarios and reasoning for determining the remedial technique.

A drinking water well has become contaminated with no warning.
With a well already contaminated, there is no time for a full RI/FS, design, and construction. Remediation in this emergency case would be called an "interim remedial action" while an investigation is performed. Because the well is already contaminated, the population will have to be provided an alternate means of drinking water. In the meantime, a relatively quick interim remedial action is to implement aboveground treatment of the contaminated water. A full RI/FS should be conducted to prevent contamination of other wells and determine the source of contamination.

Following a detailed remedial investigation (RI), a drinking water well is expected to become contaminated in 5 years.
To protect the drinking water well, pump-and-treat must be used. Groundwater extraction wells should be installed at a location that provides hydraulic containment of the plume, which at this point is located far from the well. The extracted groundwater should be treated using a technology appropriate for the COC.

A tanker truck rolls over on a highway and spills a chemical on soil adjacent to a sensitive lake.
Before the chemical reaches the lake, it's important to excavate the contaminated soil as quickly as possible and dispose of it appropriately.

Students and teachers in an elementary school notice a chemical odor and experience headaches.
Because of the emergency nature of this scenario, the school should first be evacuated. A short-term remedial solution is to install vapor wells and build a soil vapor extraction system to extract the vapors from the vadose zone under the school. This process can take several weeks at least. In the meantime, a full RI/FS should be conducted.

9.3.2 Treat the Source of Contamination

A source of contamination can be discovered in many ways. A leaking underground storage tank (LUST) at a chemical plant can be discovered after several days, or even years, of record-keeping showing that the tank has less fuel than expected. In other cases, a source of contamination is only discovered after people or the environment have suffered from its effects. The scenarios below are the same as the ones from the previous section, where now the source of contamination is known. We offer reasoning for handling the source of contamination.

A drinking water well has become contaminated with no warning. Engineers review the fueling records from a nearby gas station and discover that one UST has a slow leak that went undetected for several years.
Excavating around the leaking UST reveals a crack in the UST and fuel in the surrounding soil. The UST should be removed, and contaminated

soil excavated. A new tank with secondary containment and a leak detection system (see Figure 2.2) should be installed, with clean soil as backfill. Groundwater in the vicinity of the tank is highly contaminated and can be treated with a combination of *in situ* bioremediation and pump-and-treat.

Following a detailed remedial investigation (RI), a drinking water well is expected to become contaminated in 5 years. Groundwater contamination is discovered to be coming from landfill leachate

The leachate collection system of the landfill should be repaired. As for the soil and groundwater contamination near the landfill source, soil vapor extraction (SVE) combined with air sparging (AS) will treat volatile organic compounds (VOCs). Other types of chemicals can be treated from the groundwater using pump-and-treat, with extraction wells in the "hot spots" providing hydraulic containment to prevent the plume from moving farther. Because contamination of the downgradient well is not imminent, slower but less-expensive technologies can be used, such as *in situ* bioremediation or monitored natural attenuation.

A tanker truck rolls over on a highway and spills a chemical on soil adjacent to a sensitive lake.

Before the chemical reaches the lake, it's important to excavate the contaminated soil as quickly as possible and dispose of it appropriately. This scenario was also shown in the previous section, and the same technique treats the source and prevents contamination of a sensitive receptor at the same time.

Students and teachers in an elementary school notice a chemical odor and experience headaches. An RI discovers that the source is a crack in the liner of a waste pond from a hydraulic fracturing (fracking) operation in the adjacent property.

Initially, the fracking operation should stop so that the pond can be emptied and repaired. The resulting soil and groundwater contamination can be aggressively treated with a combination of excavation, *in situ* chemical oxidation of excavated soils, and groundwater pump-and-treat.

9.3.3 MONITOR THE ACHIEVEMENT OF REMEDIATION GOALS

Monitoring whether the remedial action is achieving its goals is a long-term process, described in Chapter 8. The goals to be achieved can be classified into two categories: (1) COC removal from its unwanted location and (2) COC effectively removed using the treatment technology.

COC Removal

Removal of the COC from soil and groundwater is performed by various technologies described in Chapters 5 and 6, such as excavation, groundwater extraction (the "pump" part of pump-and-treat), and *in situ* bioremediation. Determining whether a COC is being removed from those areas is done by the techniques described in Chapter 8, which focus on monitoring whether the COC is decreasing or absent from the soil and groundwater.

COC Treatment

While COCs may be removed from their unwanted locations underground, they often are treated aboveground. Measures need to be taken to ensure that those aboveground, *ex situ* systems are effectively treating COCs so that contamination does not occur elsewhere. Typical COC treatment is done using groundwater treatment (the "treat" part of pump-and-treat, Chapter 6) and off-gas treatment (Chapter 7). Monitoring of the effectiveness of these treatment systems is described in Section 8.4.3 for groundwater treatment and Section 8.2.1 for off-gas treatment.

9.3.4 PERIODICALLY OPTIMIZE THE REMEDIAL APPROACH

After a remedial alternative is selected through the FS process and constructed, monitoring and O&M continue for months to years and sometimes decades. The long-term duration of remediation approaches requires that the strategy be re-evaluated and optimized periodically. The re-evaluation may result in the choice of a new remedial strategy that meets the remediation goals while decreasing time, cost, and effort to complete the site remediation. This is called remediation process optimization (RPO). Some documents that describe RPO in detail are *Remediation Process Optimization: Identifying Opportunities for Enhanced and More Efficient Site Remediation* by the Interstate Technology Regulatory Council and *Guidance for Optimizing Remedial Action Operation* by NAVFAC and Battelle for the U.S. Navy, which administers many remediation sites at its bases.

The time, cost, and effort of remediation can be illustrated by Figure 9.2. In this figure, we can see that long-term monitoring and O&M last for many years

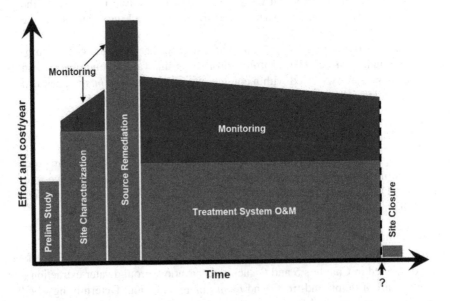

FIGURE 9.2 Typical plot of remediation effort and cost vs. time. (Source: ITRC 2004.)

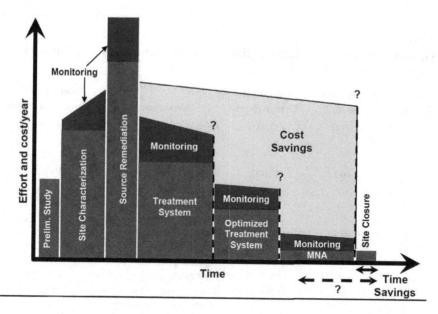

FIGURE 9.3 Plot of optimized remediation effort and cost vs. time, showing where cost savings can occur. (Source: ITRC 2004.)

and costs overall much more than the initial assessment and remediation system construction. The benefits of RPO can be seen in Figure 9.3, where monitoring and O&M cost and effort, and even the time to completion, decrease as the site's subsurface becomes better understood and remediation systems can be optimized.

The types of activities in RPO can be seen in Table 9.1. This table outlines options for optimizing a pump-and-treat system (127 gallons per minute (gpm) for seven wells) for hydraulic containment and aquifer restoration of a site contaminated with benzene, toluene, ethylbenzene, and xylene (BTEX). The table is based on a site in Massachusetts reported in ITRC (2004). Nine options are being considered for RPO, and the implementation cost and O&M cost change of each option are included in the second and third columns. The last two columns describe the proposed O&M activity and its potential improvement in removing COCs.

Example 9.1: Evaluating Options for Remediation Process Optimization (RPO)

Regarding the activities described in Table 9.1, if all of the options except the first one were implemented in one year, how many years would it take to recoup the cost of implementation? How much money would be saved each year after the implementation costs had been recouped?

TABLE 9.1

Options for Performing Remediation Process Optimization at a Pump-and-Treat Site.

Option	Implementation Cost ($)	O&M Cost Change ($/yr)	Description	Potential COC Removal Improvement
1	400,000	(30,000)	Replace air strippers with more efficient units	High
2	30,000	(40,000)	Improve filter performance	Medium
3	-	(144,000)	Eliminate full-time security	None
4	100,000	(1,260,000)	Automate system and reduce full-time operating staff from 10 to 2 employees	None
5	50,000	(140,000)	Remove pumps from inefficient wells and combine flows	None
6	25,000	(55,500)	Replace bioreactor with GAC	Medium
7	30,000	(30,000)	Combine discharge points	None
8	-	(55,000)	Reduce sampling frequency and number of wells sampled	None
9	750,000	(192,000)	Replace air strippers with tray aerators	High

Note: Numbers in parentheses mean a decrease in cost.

SOLUTION:

The implementation cost for options 2 through 9 would total $985,000, and $1,916,500 would be saved each year.

It would take 985,000/1,916,500 = 0.51 years, or approximately 6 months to recoup the implementation costs.

After the second full year (so from the third year on) of O&M using RPO, an annual amount of $1,916,500 would be saved from the third year on.

Discussion: The analysis illustrated in this example is made solely on costs, but consideration must be given to other factors. For example, the neighboring community should be consulted regarding Option 3, eliminating full-time security.

9.4   COST CONSIDERATIONS

Evaluating several remedial alternatives for cleaning up a site almost always involves a life cycle cost assessment (LCCA), which in effect is the net present worth (NPW), or net present value (NPV), of the life of the project. Here, we provide a few fundamental principles for calculating the LCCA for remedial alternatives. A more detailed approach to this topic can be found in the U.S. EPA and U.S. Army Corps of Engineers' document *A Guide to Developing and Documenting Cost Estimates During the Feasibility Study* (U.S. EPA and U.S. ACE 2000).

9.4.1 CAPITAL COSTS AND OPERATION AND MAINTENANCE (O&M) COSTS

Capital costs are the initial costs for the design, construction, and equipment needed to operate a remediation system. Operational costs include monitoring, operation, maintenance, and repair costs that are required to keep the project in operation, for example, costs of energy and labor for maintenance.

A more detailed look at the cost items for capital and O&M costs is included in Table 9.2, for a site using air sparging (AS) and soil vapor extraction (SVE) for groundwater and vadose zone soil remediation.

9.4.2 LIFE CYCLE COST ASSESSMENT (LCCA)

Costs that are considered in a typical LCCA for remediation are capital and O&M costs. The numerical value of life-cycle costs (LCC) is typically defined as shown in Eq. 9.1.

$$LCC = Capital\ Cost + NPW_{O\&M} \qquad (9.1)$$

Where $NPW_{O\&M}$ is the net present worth of O&M costs. Because remediation projects last many years, an engineering economic analysis is necessary. In engineering economic analysis, we consider that the capital cost is incurred in Year 1, and the O&M costs are incurred in Years 1 through n. Because the project owner is likely to have money invested to pay for the remediation, then we must calculate the net present worth (NPW) of the annualized O&M costs. An interest rate, i, is used to adjust for the expected growth of the investment over the years.

Therefore, for the number of years, n, that the project will operate, the net present worth cost considering equal annualized O&M costs is

$$NPW_{O\&M} = A \times \frac{(i+1)^n - 1}{i(1+i)^n} \qquad (9.2)$$

where A is the projected annual O&M expense, and i is the projected interest rate for n years. Note that $NPW_{O\&M}$ is not merely $A \times n$ because when money is invested and grows at a rate of i, less money is needed to be available at the outset of the project. Therefore, the term $\dfrac{(i+1)^n - 1}{i(1+i)^n}$ is a number less than n.

TABLE 9.2

Cost Items for an Air Sparging and Soil Vapor Extraction Project. (Based on U.S. EPA and U.S. ACE 2000)

Capital Cost Items	O&M Cost Items

Professional/Technical Services

- Project Management – planning, community relations, bid and contract administration, permitting
- Remedial Design – analysis of field data, pilot studies, plans and specifications, construction cost estimate
- Construction Management – construction observation, schedule tracking, change order review, design modifications

- Project Management – planning, community relations, cost and performance reporting, permitting
- Technical Support – O&M manual updates, O&M oversight, progress reports

Institutional Controls

- Zoning, property easements, deed notice

- Zoning, property easements, deed notice

Field-Related Activities

- Mobilization/Demobilization – construction equipment, implementation plans (health and safety, stormwater pollution prevention plan, etc.), temporary facilities (office trailer, storage, fencing), as-built drawings, O&M manual
- Sampling – air monitoring, health and safety monitoring, personal protective equipment, soil sampling, groundwater sampling, laboratory chemical analysis, geotechnical testing
- Site Work – clearing and grubbing, excavation, stockpiling, hauling, waste disposal, grading, utilities (electrical, telephone, water/gas), storm drainage
- Air Sparging – purchase and installation of injection wells, air compressor, piping
- Soil Vapor Extraction – purchase and installation of extraction wells, blower, treatment system, piping

- Monitoring, sampling, testing, and analysis – air monitoring, health and safety monitoring, personal protective equipment, soil sampling, groundwater sampling, process water sampling, process air sampling, laboratory chemical analysis, chemical data management
- Remediation Systems – operations labor, maintenance labor, equipment upgrade/replacement/repair, spare parts, equipment rental, consumable supplies, utilities
- Disposal – waste disposal, fees

A common short-hand representation of the factor on the right side of Eq. 9.2 is the term (P/A, i%, n):

$$\left(P \, / \, A, i\%, n \right) = \frac{\left(i + 1 \right)^{n} - 1}{i \left(1 + i \right)^{n}} \tag{9.3}$$

Therefore, $NPW_{O\&M}$ can also be written as

$$NPW_{O\&M} = A \times \left(P \, / \, A, i\%, n \right) \tag{9.4}$$

Because the term (P/A, $i\%$, n) is less than n years, the LCC is a value less than the capital cost plus the O&M cost simply multiplied by n. The LCC when considering $NPW_{O\&M}$ then represents how much the payer must invest (in a bond, for example) to ensure that cash for the annual costs is available. If the initial investment increases by the predicted interest rate i, then there should be enough cash available to pay for the remediation project in the long-term, for n years. The next example demonstrates how this calculation is useful when comparing remedial alternatives.

Example 9.2: Comparing the Net Present Worth of Two Remedial Alternatives

Two treatment alternatives are being considered to treat the off-gas from a soil vapor extraction (SVE) system extracting VOCs from the vadose zone at a contaminated site. Calculate the life cycle cost (LCC) of each and state which one is less costly. The system is expected to operate for 15 years, and the predicted interest rate is 6%.

	Alternative A	Alternative B
Capital cost	$950,000	$600,000
Annual O&M cost (A)	$200,000	$300,000

SOLUTION:

(a) Calculate the $NPW_{O\&M}$ for each alternative:

$$NPW_{O\&M,A} = A_A \times (P/A, 6\%, 15) = 200,000 \times \frac{(0.06+1)^{15}-1}{0.06(1+0.06)^{15}}$$

$$= 200,000 \times 9.7122 = \$1,942,400$$

$$NPW_{O\&M,B} = A_B \times (P/A, 6\%, 15) = 300,000 \times \frac{(0.06+1)^{15}-1}{0.06(1+0.06)^{15}}$$

$$= 300,000 \times 9.7122 = \$2,913,700$$

(b) Calculate the LCC of each alternative:

$$LCC_A = Capital_A + NPW_{O\&M,A} = \$950,000 + \$1,942,400 = \$2,892,400$$

$$LCC_B = Capital_B + NPW_{O\&M,B} = \$600,000 + \$2,913,700 = \$3,515,700$$

Alternative A is less costly, even though the capital cost is higher.
Discussion: The LCC values are less than the sum of the capital cost plus O&M cost multiplied by 15 years. This is because less cash is needed to be invested upfront if it will grow at an interest rate of i.

9.5   ENVIRONMENTAL JUSTICE

Remediating contaminated sites seems to automatically improve community well-being because of achieving the goal of protecting human health and the environment. But the reality can be different from that.

It is widely known that contaminated sites throughout the world, whether active industrial facilities or hazardous waste disposal sites, are located in or near low-income and minority communities. Many research studies in the fields of public health, public policy, sociology, geography, and economics report this problem (Bryson 2012; Dowling et al. 2015; Pasetto et al. 2019). As a result of the proximity of low-income and minority neighborhoods to these sites, these populations suffer from environmental injustice and become more susceptible to the harmful health effects of pollutants.

In the United States, in response to this, the EPA created the Office of Environmental Justice in 1992 (https://www.epa.gov/environmentaljustice). The EJ Office provides grants and resources, forms partnerships with communities, and adds the involvement of community concerns to existing environmental planning frameworks. For example, the report *Promising Practices for EJ Methodologies in NEPA Reviews* (Federal Interagency Working Group on Environmental Justice 2016) describes effective practices to improve environmental justice issues in the processes under the jurisdiction of the National Environmental Policy Act (NEPA). The Office has also developed a mapping tool, EJSCREEN, to identify demographic data and proximity to possible environmental hazards, such as hazardous waste sites, Superfund sites, and wastewater discharges. This geographic information system (GIS)-based tool can be found and used freely online at https://ejscreen.epa.gov/mapper/.

Environmental justice researchers who examine contaminated sites often do so in the context of brownfields, as defined in Chapter 3. A brownfield is a property in which its expansion, redevelopment, or reuse may be complicated by the presence or potential presence of a hazardous substance/pollutant/contaminant. Brownfield redevelopment, while cleaning the environment and protecting human health, can also lead to gentrification, tying environmental justice issues to socioeconomic issues (Bryson 2012). Gentrification is the redevelopment of urban areas for improved aesthetics and increased housing prices, with the consequences of attracting affluent residents but driving out low-income residents. Even decisions on which brownfields to rehabilitate have environmental justice implications. For example, when considering only potential economic earnings when deciding on which brownfield site to rehabilitate, those sites in disadvantaged communities may be left out (McCarthy 2009).

Remediation professionals are often required to speak at public hearings and prioritize sites to be remediated. Although an engineer and scientist's focus is traditionally on the technical aspects of the project, these professions, often by their codes of ethics, are called upon to act on behalf of populations. For example, the American Society of Civil Engineers' (ASCE) Canon 1 of the Code of Ethics states "Engineers shall hold paramount the safety, health and welfare of the public and shall strive to comply with the principles of sustainable development in the performance of their professional duties." Canon 8 of the same Code of Ethics states "Engineers shall, in all matters related to their profession, treat all persons fairly and encourage equitable participation without regard to gender or gender identity, race, national origin, ethnicity, religion, age, sexual orientation, disability, political affiliation, or family, marital, or economic status" (ASCE 2020).

9.6 SUSTAINABLE REMEDIATION

At this point, it may be evident that environmental remediation projects have to balance the protection of human health and the environment, optimize cost and effort, and be equitable to disadvantaged populations. This balance is tenuous but something to strive for. The term 'sustainable remediation' refers to the intentional implementation of a remedial action beyond the typical RI/FS process that considers environmental protection, economic prosperity, and social equity.

9.6.1 SUSTAINABILITY

The term 'sustainability' came to the forefront in 1987 in the context of sustainable development for nations. The term 'sustainable development' was popularized from the writing of Norway's Prime Minister Gro Harlem Bruntland (WCED 1987):

> *Sustainable development is development that meets the needs of the present without compromising the ability of future generations to meet their own needs. It contains within it two key concepts:*
>
> * *the concept of 'needs,' in particular the essential needs of the world's poor, to which overriding priority should be given; and*
> * *the idea of limitations imposed by the state of technology and social organization on the environment's ability to meet present and future needs.*

Since then, numerous publications, too many to name here, have explored how the concept of sustainability can be applied to all aspects of life. The concept of the Triple Bottom Line, or three pillars of sustainability, was introduced by John Elkington in his book *Cannibals with Forks: Triple Bottom Line of 21st Century Business* (Elkington 1999). This book took sustainability beyond something "nice to do" and showed how businesses could be leaders in Triple Bottom Line accounting: economic prosperity, environmental protection, and social equity. These three pillars are typically visualized as shown in Figure 9.4, with three overlapping sets of a Venn diagram. Each set represents a pillar, and the middle overlap represents the achievement of sustainability.

Many facets of society have embraced the concept of sustainability, and many professional societies related to environmental assessment and remediation have sustainability statements, for example:

* American Institute of Chemical Engineers' (AIChE) Climate Change Policy Statement:

  Our community is committed to playing a leadership role in offering solutions to climate change through systems and other approaches that will create resilient and sustainable processes, products, and facilities. AIChE and its members

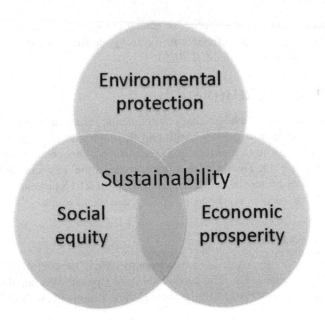

FIGURE 9.4 The three pillars of sustainability.

hold paramount the present and future safety, health and welfare of the public, and the protection of the environment in performance of their professional duties (AIChE 2019).

- American Society of Civil Engineers (ASCE) Policy Statement 418:

  The American Society of Civil Engineers (ASCE) defines sustainability as a set of economic, environmental and social conditions (aka "The Triple Bottom Line") in which all of society has the capacity and opportunity to maintain and improve its quality of life indefinitely without degrading the quantity, quality or the availability of economic, environmental and social resources. Sustainable development is the application of these resources to enhance the safety, welfare, and quality of life for all of society (ASCE 2018a).

The ASCE statement very clearly encompasses the Triple Bottom Line. ASCE goes further in its Policy Statement 556 to recommend project owners, who hire civil engineers, to "incorporate sustainability principles and practices in the development of infrastructure projects. ASCE supports the continued education and outreach to owners on the positive impacts and importance of sustainable infrastructure projects" (ASCE 2018b).

To promote sustainability in infrastructure (e.g., bridges, highways, wastewater treatment plants, power generation facilities), the Envision™ infrastructure sustainability rating tool (www.sustainableinfrastructure.org) was created by ASCE, the American Council of Engineering Companies (ACEC), and the American Public Works Association (APWA). Infrastructure projects that are planned using

Envision™ must meet many "credits" that are sustainability indicators addressing the Triple Bottom Line. Some of these credits include:

- Reclaiming brownfields,
- Enhancing public health and safety,
- Fostering collaboration and teamwork,
- Planning for long-term monitoring and maintenance,
- Reducing greenhouse gas emissions,
- Reducing construction waste, and
- Preserving water resources.

This small sampling of the 64 total available credits apply to remediation projects. Therefore, remediation projects in and of themselves can follow the Envision™ guidelines and meet its sustainability credits.

We next look at how the Triple Bottom Line is applied in sustainable remediation.

9.6.2 SUSTAINABILITY IN REMEDIATION PROJECTS

Sustainable remediation can be defined as remediation that addresses the sustainability triple bottom line. A term that is commonly used with sustainable remediation is green remediation. This remediation focuses on the environmental aspect of sustainability, mostly concerned with lowering the environmental footprint of a remedial action.

Several publications address sustainable remediation in more detail and are good follow-up readings after the reader is comfortable with the many technologies used in remediation.

- *Green and Sustainable Remediation: A Practical Framework* (ITRC 2011a);
- *Green and Sustainable Remediation: State of the Science and Practice* (ITRC 2011b);
- *Green Remediation: Incorporating Sustainable Environmental Practices into Remediation of Contaminated Sites* (U.S. EPA 2008); and
- *Sustainable Remediation of Contaminated Sites* (Reddy and Adams 2015).

An easily accessible tool for evaluating remediation sustainability is the Sustainable Remediation Forum (SURF) Metric Toolbox for evaluating sustainability during the life of a site investigation and remediation project (Butler et al. 2011). This toolbox can be found at https://www.sustainableremediation.org/guidance-tools-and-other-resources and is composed of a series of tables for several stages of the remediation process: investigation, remedy selection, design, and construction. Each table contains descriptions of objectives, metrics, implementation guidance, benefits, and challenges for implementing sustainable parameters.

Environmental Sustainability

Environmental sustainability goes beyond the environmental objectives for the contaminated site and fosters a decrease in the environmental footprint of the remedial action. For example, reducing the environmental footprint would decrease emissions of greenhouse gases and the acquisition of raw materials for the remediation strategy.

A common way of tracking the environmental footprint of an activity is to conduct an environmental life cycle assessment (LCA) (U.S. EPA and SAIC 2006). An LCA requires four stages of activities:

- Goal definition and scoping,
- Inventory analysis,
- Impact assessment, and
- Interpretation of results.

LCA is a data-driven and data-intensive process and requires computer tools and data inventories. A complete LCA quantifies the environmental impacts on the following categories:

- Global warming by analyzing data on greenhouse gas emissions;
- Stratospheric ozone depletion by analyzing data on ozone-depleting chemicals;
- Acidification by analyzing data on the release of sulfur and nitrogen oxides (SO_x and NO_x);
- Eutrophication by analyzing data on the release of nitrogen and phosphorus compounds;
- Photochemical smog by analyzing data on non-methane hydrocarbons;
- Terrestrial toxicity by analyzing data on the release of chemicals to rodents (*a remediation project in and of itself would reduce this impact but potentially cause other impacts on this list*);
- Aquatic toxicity by analyzing data on the release of chemicals toxic to fish (*a remediation project in and of itself would reduce this impact but potentially cause other impacts on this list*);
- Human health by analyzing data on the release of chemicals toxic to humans and released to air, water, and soil (*a remediation project in and of itself would reduce this impact but potentially cause other impacts on this list*);
- Resource depletion by analyzing data on minerals and fossil fuels used
- Land use by analyzing data on wastes disposed to landfills; and
- Water use by analyzing data on water used or consumed.

The takeaway message here is that cleaning up the environment can cause adverse effects on the environment. For example, remediating a site rids it of toxic chemicals but may contribute to more greenhouse gases. The EPA conducted a study on all National Priority List (NPL) sites using the five remediation technologies listed in Table 9.3 and found that from 2008 to 2030 (data were extrapolated) those sites are estimated to spend over $1.4 billion on electricity and emit over 9 million metric tons of CO_2, the equivalent of operating two coal-fired power plants for one year

TABLE 9.3

Estimated CO_2 emissions from remediation technologies at NPL sites using five technologies over 23 years (U.S. EPA 2008)

Technology	Estimated CO_2 Emissions Annual Average (Metric Tons)	Total Estimated CO_2 Emissions in 2008–2030 (Metric Tons)
Pump & Treat	323,456	7,439,480
Thermal Desorption	57,756	1,328,389
Multi-Phase Extraction	12,000	276,004
Air Sparging	6,499	149,476
Soil Vapor Extraction	4,700	108,094
Total	404,411	9,301,443

(U.S. EPA 2008). If reducing CO_2 emissions is a priority, then one way to practice environmentally sustainable remediation would be to consider using more passive techniques, such as enhanced bioremediation, phytoremediation, monitored natural attenuation (MNA), and engineered wetlands (which combine processes of bioremediation, phytoremediation, and MNA) as long as they meet the required remediation goals.

It may come as no surprise that remediation process optimization (RPO) covered in Chapter 8 overlaps with concepts of sustainable remediation. For example, the United States Navy conducted a study on remediation sites at its bases within the context of optimization and concluded that after a certain amount of contaminant mass is removed, the other environmental impacts caused by the remediation system (e.g., emissions of criteria air pollutants and greenhouse gases (GHG), changes in land and ecosystem, and impact to the community and workers due to noise and safety) were not worth additional reduction in contaminant mass (Figure 9.5) (NAVFAC Optimization Workgroup and Battelle Memorial Institute 2012).

Social Sustainability

Many tools and methodologies are available for remediation professionals to track the societal impacts of their projects (Harclerode et al. 2011). Although not used consistently throughout the remediation profession, the leading social indicator categories can be summarized as:

- **Health and safety** of remediation workers and the neighboring community;
- **Economic vitality** of the community through the hiring of local workers, outreach activities, and land redevelopment;
- **Stakeholder collaboration** so that project workers can detect unintended consequences before they happen;

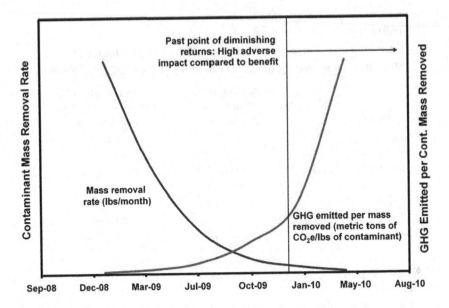

FIGURE 9.5 Data visualization to detect adverse impacts on the environment by remediation systems. (Source: NAVFAC Optimization Workgroup and Battelle Memorial Institute 2012.)

- **Benefits to the community at large** by improving the area for the population's activities;
- **Undesirable community impact minimized**, such as traffic, noise, and dust;
- **Social justice** by continuing to make affordable housing available, providing jobs to local residents, and provide new, clean gathering spaces;
- **Value of ecosystem services and natural resources capital** re-established by cleaning contamination and restoring ecosystems and water features;
- **Risk-based land management and remedial solutions** to address community-specific health concerns;
- **Regional and global societal impacts** alleviated by performing sustainable remediation that reduces negative effects on public health and climate; and
- **Contribution to local and regional sustainability policies and initiatives** by performing state-of-the-practice sustainability activities, such as using renewable energy sources, employing local residents, and performing health risk analysis specific to the local community.

Something that becomes apparent after studying the social indicators above is the overlap among social, environmental, and economic indicators. However, the overlap does not mean one can ignore one or more of the three pillars of sustainable remediation. Sustainable projects require an intentional pursuit and demonstration of the achievement of all three pillars.

Economic Sustainability

Economic sustainability in remediation consists of projects that generally can optimize cost expenditures and also benefit the local community economically. The SURF Metric Toolbox (SURF 2020) cites the following indicators for economic sustainability in a range of site activities.

Remedy Selection (Feasibility Study phase)

- Create jobs in the community, measured by new space for commercial development or recreation;
- Select minimum essential materials;
- Minimize the timing to achieve future land use; and
- Minimize off-site disposal of solid waste.

Remedial Design

- Optimize the design to reduce the number of wells, pumps, and chemicals;
- Conduct optimization designs throughout the life of the project;
- Maximize the future land use and area of the project;
- Increase local jobs based on redevelopment of the area;
- Minimize and consolidate transportation to and from the site to reduce fuel costs;
- Design for minimum essential materials for construction; and
- Use on-site renewable energy to reduce operational costs.

Remedial Construction

- Minimize and consolidate transportation to and from the site to reduce fuel costs;
- Ensure the correct equipment size for tasks to balance cost and efficiency;
- Minimize idling time of equipment;
- Reuse site construction materials where possible;
- Purchase from vendors that have sustainable programs and policies;
- Use energy-efficient systems and equipment for field buildings; and
- Reduce on-site risk of injury or accidents.

Economic sustainability indicators are common-sense good practices when an environmental professional is intentional in pursuing sustainability. This requires good planning, a well-functioning project team, and good relationships with stakeholders, including construction contractors, the project owner, government oversight personnel, and the local community.

9.7   SUMMARY

This chapter referred to the earlier chapters of the book and showed how they come together as a whole. In addition, this chapter addresses modern practices in sustainability to promote efficient and equitable remedial actions. Topics covered were:

- Feasibility study,
- Remediation strategies,
- Cost considerations,
- Environmental justice, and
- Sustainable remediation.

9.8   PROBLEMS AND ACTIVITIES

9.1.   Refer to the site conceptual model (SCM) in Figure 4.1 in Chapter 4 and write a paragraph proposing a remediation strategy that will prevent the contamination from reaching sensitive receptors and treat the source.

9.2.   Use Table 9.1 to answer this question. Assume that the current operation and maintenance cost is $3,000,000 per year. The site supervisor has a budget of $80,000 to invest in optimizing the remediation process. Knowing that people who live near the site are very concerned about the contamination and are firmly against both reducing the full-time security and reducing the sampling frequency and number of wells sampled, which options can be utilized to save the most money?

9.3.   Use Table 9.1 to answer this question. Assume that the mass of COC removal has reached an asymptote with respect to time with the current treatment process. The site supervisor has a budget of $150,000. Knowing that the COC removal efficiency has to be increased, which options can be utilized to save the most money?

9.4.   The graph (Figure 9.6) shows the projection of the cumulative savings per year for eight years for two remediation options that both increase the efficiency of remediation at a site. Which option should be chosen if the project is expected to continue until Year 3? Until Year 4?

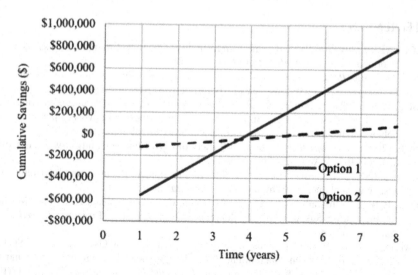

FIGURE 9.6   Graph of cost vs time of two remedial options (Problem 9.4).

9.5.   An engineer is working on a feasibility study for a groundwater pump-and-treat system using activated carbon for treatment. The estimated costs are shown in the tables below. The life of the project is expected to last 20 years at a 5% interest rate. Calculate the total life cycle cost.

Capital cost item	Cost ($)
Engineering design	40,000
Well drilling and installation	120,000
Concrete pad construction	40,000
Purchase and installation of GAC vessels	75,000
Trenching and laying pipe	60,000
Purchase and installation of control system	30,000

Annual O&M item	Cost ($/year)
Electricity	4,000
GAC replacement and disposal	50,000
O&M labor, including data analysis and reporting	190,000
Analytical samples	22,000

REFERENCES

AIChE. (2019). *Institute Policies*. American Institute of Chemical Engineers, https://www.aic he.org/about/governance/policies (Aug. 5, 2020).

ASCE. (2018a). *Policy Statement 418: The Role of the Civil Engineer in Sustainable Development*. American Society of Civil Engineers, Reston, Virginia.

ASCE. (2018b). *Policy Statement 556: Owners Commitment to Sustainability*. American Society of Civil Engineers, https://www.asce.org/issues-and-advocacy/public-policy/ policy-statement-556---owners-commitment-to-sustainability/ (Aug. 6, 2020).

ASCE. (2020). *Code of Ethics*. American Society of Civil Engineers, https://www.asce.org/ code-of-ethics/ (Aug. 5, 2020).

Bryson, J. (2012). "Brownfields Gentrification: Redevelopment Planning and Environmental Justice in Spokane, Washington." *Environmental Justice*, 5(1), 26–31.

Butler, P. B., Larsen-Hallock, L., Lewis, R., Glenn, C., and Armstead, R. (2011). "Metrics for Integrating Sustainability Evaluations Into Remediation Projects." *Remediation Journal*, 21(3), 81–87.

Dowling, R., Ericson, B., Caravanos, J., Grigsby, P., and Amoyaw-Osei, Y. (2015). "Spatial Associations Between Contaminated Land and Socio Demographics in Ghana." *International Journal of Environmental Research and Public Health*, 12(10), 13587–13601.

Elkington, J. (1999). *Cannibals with Forks: Triple Bottom Line of 21st Century Business*. John Wiley & Sons, New York.

Federal Interagency Working Group on Environmental Justice. (2016). *Promising Practices for EJ Methodologies in NEPA Reviews*. EPA 300B16001. U.S. Environmental Protection Agency, Washington, D.C.

Harclerode, M., Risdsdale, D. R., Darmendrail, D., Bardos, P., Alexandrescu, F., Nathanail, P., Pizzol, L., and Rizzo, E. (2011). "Integrating the Social Dimension in Remediation Decision-Making: State of the Practice and Way Forward." *Remediation Journal*, 26(1), 11–42.

ITRC. (2004). *Remediation Process Optimization: Identifying Opportunities for Enhanced and More Efficient Site Remediation*. RPO-1, Interstate Technology & Regulatory Council, Remediation Process Optimization Team, Washington, DC.

ITRC. (2011a). *Green and Sustainable Remediation : A Practical Framework*. Interstate Technology & Regulatory Council, Green and Sustainable Remediation Team, Washington, DC.

ITRC. (2011b). *Green and Sustainable Remediation: State of the Science and Practice*. Interstate Technology & Regulatory Council, Green and Sustainable Remediation Team, Washington, DC.

McCarthy, L. (2009). "Off the Mark?: Efficiency in Targeting the Most Marketable Sites Rather Than Equity in Public Assistance for Brownfield Redevelopment." *Economic Development Quarterly*, 23(3), 211–228.

NAVFAC Optimization Workgroup, and Battelle Memorial Institute. (2012). *Guidance for Optimizing Remedial Action Operation (RA-O)*. UG-NAVFAC EXWC-EV-1301. U.S. Navy, Naval Facilities Engineering Command, Columbus, OH.

Pasetto, R., Mattioli, B., and Marsili, D. (2019). "Environmental Justice in Industrially Contaminated Sites. A Review of Scientific Evidence in the WHO European Region." *International Journal of Environmental Research and Public Health*, 16(6).

Reddy, K., and Adams, J. (2015). *Sustainable Remediation of Contaminated Sites*. Momentum Press, New York.

SURF. (2020). *Guidance, Tools, and Other Resources.* Sustainable Remediation Forum, https://www.sustainableremediation.org/guidance-tools-and-other-resources (Aug. 6, 2020).

U.S. EPA. (1988). *Guidance for Conducting Remedial Investigations and Feasibility Studies Under CERCLA.* EPA/540/G-89/004, U.S. Environmental Protection Agency Office of Emergency and Remedial Response and U.S. Army Corps of Engineers Hazardous, Toxic, and Radioactive Waste Center of Expertise, Washington, DC.

U.S. EPA. (1990). *The Feasibility Study : Detailed Analysis of Remedial Action Alternatives. 9355.3–01F.* U.S. Environmental Protection Agency Office of Solid Waste and Emergency Response, Washington, DC.

U.S. EPA. (2008). *Green Remediation: Incorporating Sustainable Environmental Practices Into Remediation of Contaminated Sites.* U.S. Environmental Protection Agency Office of Solid Waste and Emergency Response, Cincinnati, OH.

U.S. EPA, and SAIC. (2006). *Life Cycle Assessment: Principles and Practice.* EPA/600/R-06/060. U.S. Environmental Protection Agency National Risk Management Research Laboratory, Cincinnati, OH.

U.S. EPA, and U.S. ACE. (2000). *A Guide to Developing and Documenting Cost Estimates During the Feasibility Study.* EPA 540-R-00-002. U.S. Environmental Protection Agency Office of Emergency and Remedial Response and U.S. Army Corps of Engineers Hazardous, Toxic, and Radioactive Waste Center of Expertise. Washington, DC.

WCED. (1987). *Our Common Future.* Oxford University Press, New York.

Index

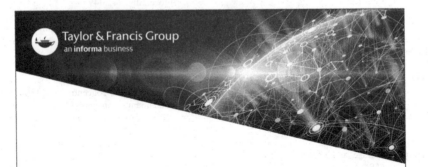

Printed in the United States
by Baker & Taylor Publisher Services